QSAR AND SPECTRAL-SAR IN COMPUTATIONAL ECOTOXICOLOGY

QSAR AND SPECTRAL-SAR IN COMPUTATIONAL ECOTOXICOLOGY

Edited By

Mihai V. Putz, PhD

Associate Professor of Theoretical Physical Chemistry,
Laboratory of Structural and Computational Physical Chemistry,
Biology-Chemistry Department, West University of Timisoara, Romania

Apple Academic Press

TORONTO   NEW JERSEY

© 2013 by
Apple Academic Press Inc.
3333 Mistwell Crescent
Oakville, ON L6L 0A2
Canada

Apple Academic Press Inc.
1613 Beaver Dam Road, Suite # 104
Point Pleasant, NJ 08742
USA

First issued in paperback 2021

Exclusive worldwide distribution by CRC Press, a Taylor & Francis Group

ISBN 13: 978-1-77463-202-4 (pbk)
ISBN 13: 978-1-926895-13-0 (hbk)

Library of Congress Control Number: 2012935658

Library and Archives Canada Cataloguing in Publication

QSAR and SPECTRAL-SAR in computational ecotoxicology/edited by Mihai V. Putz.

Includes bibliographical references and index.
ISBN 978-1-926895-13-0
1. Environmental toxicology–Computer simulation. 2. Environmental toxicology–Mathematical models.
3. Structure-activity relationships (Biochemistry). I. Putz, Mihai V

RA1226.Q73 2012   615.9›020113   C2011-908705-7

Contents

List of Contributors

Sergiu Andrei Chicu
Siegstr. 4, Köln, D-50859, Germany.

Adrian Chiriac
Biology-Chemistry Department, West University of Timişoara, Pestalozzi Street No.16, Timişoara, RO-300115, Romania.

Corina Duda-Seiman
Laboratory of Organic Chemistry, Biology-Chemistry Department, West University of Timişoara, Pestalozzi Street No.16, Timişoara, RO-300115, Romania.

Daniel M. Duda-Seiman
University of Medicine and Pharmacy "Victor Babes" Timisoara, B-dul C.D. Loga 49, RO-300020, Romania.

Luciana Ienciu
Whatman, Part of GE Healthcare, Inc, 200 Park Avenue Suite 210, Florham Park, NJ 07932-1026, USA.

Marius Lazea
Biology-Chemistry Department, West University of Timişoara, Pestalozzi Street No.16, Timişoara, RO-300115, Romania.

Vasile Ostafe
Biology-Chemistry Department, West University of Timişoara, Pestalozzi Street No.16, Timişoara, RO-300115, Romania.

Ana-Maria Putz
Institute of Chemistry Timisoara of the Romanian Academy, 24 Mihai Viteazul Bld., Timisoara, RO-300223, Romania & Biology-Chemistry Department, West University of Timişoara, Pestalozzi Street No.16, Timişoara, RO-300115, Romania.

Mihai V. Putz
Laboratory of Structural and Computational Physical Chemistry, Biology-Chemistry Department, West University of Timişoara, Pestalozzi Street 16, Timişoara, RO-300115, Romania.

List of Abbreviations

Ach	Acetylcholine
AChE	Acetylcholinesteras
BCRP	Breast cancer resistance protein
BMIM	1-n-Butyl-3-methylimidazolium
CCD	Chemical category database
CoMFA	Comparative molecular field analysis
D.m.	*Daphnia magna*
DEA	Drug Enforcement Administration
DMPC	Dimirystoyl phosphatidyl choline
DTA	Direct toxicity assessment
ESIP	Element specific influence parameter
FDA	Fluorescein diacetate
GA	Genetic algorithms
GERD	Gastroesophageal reflux disease
H.e.	*Hydractinia echinata*
HPV	High production volume
ICES	International Council for the Exploration of the Sea
IEMAD	Inter-endpoint molecular activity difference
IEND	Inter-endpoint norm difference
IL	Ionic liquid
ISO	International Standard Organization
LD	Lethal dose
LOAEC	Low observed adverse effect concentration
L-R	Ligand-receptor
LUMO	Lowest unoccupied molecular orbital
MDR	Multidrug resistance
MTD	Minimal topological difference
MX	Mitoxantrone
NN	Neuronal-network
NTP	National Toxicology Program
OECD	Organisation for Economic Co-operation and Development
P.p.	*Pimephales promelas*
PCA	Principal component analysis
PEC	Predicted environmental concentration
PLS	Partial least squares

PNEC	Predicted no-effect concentration
POL	Polarizability
PPIs	Proton-pump inhibitors
QSARs	Quantitative structure-activity relationships
QSPRs	QSARs for membrane permeability
QSSA	Quasi-steady-state approximation
QUANTUM-SAR	Quantum nature of tuning metabolism-structure-activity relationship
REACH	Registration, evaluation, and authorization of chemicals
S-E	Substrate-enzyme
SEE	Standard error of estimate
SEM	Scanning-electron microscopy
SPECTRAL-SAR	Special computing trace of algebraic structure-activity relationship
SQ	Sum of squares
SR	Sum of the residues
T.p.	*Tetrahymena pyriformis*
TFMSi	Bis(trifluoromethylsulfonyl)imide
TGD	Technical guidance document
V.f.	*Vibrio fisheri*
VOCs	Volatile organic compounds

Preface

In the last years, the world scientific research was focused on the so called *green chemistry*, which consists in the efforts to reduce or eliminate the use or production of the dangerous substances (with toxic potential) in synthesis, main stream, and application of the chemical compounds through pre-industrial or computational design. As such on all meridians new specific organizations and laws of validation of the entered compounds in environment or everyday and medical life have raised: the first taxonomical groups emerged in United States by the *Environmental Protection Agency* (EPA, 1991) followed by the European agency *Umweltbundesamt* (1997) and by the *Environment Canada* (1999). However, at the level of European Union, since the Strategy on Management of Substances (SOMS, 2001) program the first step was made toward establishing by the European Commission, on October 23, 2003, to the *Registration, Evaluation, Authorization, and Restriction of Chemicals* (REACH) norms establishing, through its directive EC no. 1907/2006, that starting from 2009 any substance with carcinomic or mutagenic potential entering in the life-cycle through market to be made only with authorization of the *European Chemical Agency* (ECMA) at Helsinki.

In this context, the fundamental research is at its turn driven by the EU laws through the directives of the *Organization of Economical and Cooperation Development* (OECD) that already credits the QSAR (Quantitative Structure-Activity Relationship) methodology as the only and certain source of computational design for the tested compounds with bio-, eco-, and pharmaco-logical impact (OECD, 2004).

Therefore, a certain conceptual-computational analysis of a compound of a series of compounds in the view of assigning its toxicity degree naturally two levels: one addresses the atomic-molecular structure together with related quantum properties while the other envisages the correlations of these properties, for example hydrophobicity, polarizability, steric effects, and so forth, with the bio-, eco-, or pharmaco-logical observed activities. Finally, it gets out the molecular mechanistic <picture> of the reactions involved in the studied chemical–biological interaction or, with other words, of the quantum chemical strength established between the ligand (the effector or the chemical) and receptor (in the target site or organism). Still, either the structure or the quantum chemical binding aspects require the advanced studies upon them, first in a separate manner, and then combined both at the intrinsic structural level and for correlating the interaction, based on the versatility of the atomic and molecular world to generate surprisingly structures and interactions just because the quantum character involved (i.e., undulatory, thus allowing the tunneling even for the energetic inaccessible potential barriers) when forming new apparently not explicated or controllable compounds by means of macroscopic procedures.

Turning to the bio-, eco-, or pharmaco-logical structure-activity relationships the success of the QSAR methods was further certified by its consecration as the official algorithm agreed by the EU when validating new chemical substances (OECD, 2004). Still, whatever the computational procedure approached, either of that of Hansch type,

3D, decisional, or orthogonal ones, the problem of delivering the molecular interaction mechanism as a QSAR analysis result was only recently furnished (Putz et al., 2006–2010), while for the first time applied in ecotoxicology with occasion with the anionic-cationic interaction study of some ionic-liquid upon the *Vibrio fischeri* and *Daphnia magna* species. This algorithm, called SPECTRAL-SAR (SPEcial Computing TRace of ALgebraic SAR) proposes a purely algebraic rethinking of the traditional statistic QSAR, which allows, through the new concepts introduced (e.g., the orthogonal space of variables, the vectorial length of the biological activity, or the algebraic correlation factor as an intensity measure of the chemical–biological interaction) the building of an optimized chart of the molecular action pathways grounded on the *minimum spectral path principle*, $\delta[A, B] = 0$ with A and B the endpoints, within a generalized space of the action norms and correlation factors. The computational results have been spectacular since rationalizing of the preceding experimental studies, this way opening new opportunities in the ligand-substrate correlation quantification.

In these conditions, the actual modeling algorithm propose a dual approach, the molecular structure-activities correlation, offering a hierarchy in predicting and controlling of the quantum chemical bonding strength at molecular, biomolecular (enzymes with primary metabolic role), and organisms (single- and many-cellular) levels through implementation and generalization of the ligand-substrate interaction scheme in the reactivity mechanisms leading with the bio-, eco-, and pharmaco-logical effects.

However, the present SPECTRAL-SAR project collects in a single volume the edited brand new approaches of fashioned QSAR studies; this is mainly achieved through modeling the quantum-chemical viable parameters at the atomic-molecular levels and then by correlating them with the bio-, eco-, and pharmaco-logical observed activities through an optimized vectorial (essentially non-statistical) algorithm. Such research direction is approached through the use of the complete analytical SPECTRAL-SAR algorithm targeting the docking optimization procedure in quantum description of the ligand-receptor interaction (having a fitting precision at the atomic level while exhibiting temporal stability against the internal metabolic and external environmental factors relating the concerned organisms). The new algorithm is unitarily presented as a new QSAR prediction (the SPECTRAL-SAR maps) for the toxicity degree of the chemical compounds depending on their inner structure and of the imbedded environment. Overall, SPECTRAL-SAR comprises the algebraic information of the chemical–biological structures computationally filtered by the minimum path principle to provide the spectral maps of eco-, bio-, and pharmaco-logical evolution of interconnected systems.

Therefore, the actual SPECTRAL-SAR project aims to satisfy the urgent requirements in validation, prediction, and control from the society side against the demographic, economical, and environmental changes having to face an ever growing production of chemicals with toxic or cancer potential. However, being an interdisciplinary approach, implying either fundamental and applicative stages, its unfolding calls original contributions of the QSAR team leaded by the present author–editor in the Laboratory of Structural and Physical Chemistry developed within the Chemistry Department of the West University of Timişoara in the last five years with applications on natural complex system analysis in either condensed, gaseous, liquid, *in vitro* or *in vivo* states, associated with different levels of manifestations of the researched chemicals by means of their structural strength in describing, understanding, and predicting of their quantum chemical bonding interacting various biological species. Overall, the present volume, through collecting an ordered series of related SPECTRAL-SAR internationally indexed papers eventually updated and completed specially for the present publishing venture by the author–editor, gives just an input for the future enterprises of intense conceptual-computationally and experimentally research aiming in profoundly studying and designing the particular systems involved in the bio-, eco-, and pharmaco-logical actions: supra-molecules of enzyme types, ionic liquids, antagonists, and inhibitors.

The book is intended to academia and research institutes and industry, at the postgraduate level (master, doctoral, and post-doctoral education) as it is currently lectured and referred in author–editor Alma Mater. All chapters "tell the story" of "SPECTRAL-SAR" method in various forms and application levels, with the unique feature being all driven by the author–editor of the book leading various colleagues in their specialization. However, although mainly as reprinted volume, many of the present chapters were revised and extended with appropriate sections and Appendices such that a unitary perspective of the role and efficacy of the method in computational ecotoxicology to be better revealed. Such updates were supported by special research program of Romanian Research Agency CNCSIS-UEFISCSU (actual CNCS-UEFISCDI) by project number PN II-RU TE16/2010-2013. Finally, the present volume would not be possible without the scientific appreciation of Profs. Eduardo A. Castro and Akbar K. Haghi, as well as by the professional assistance of Ms. Sandra Jones Sickels (Marketing Director at Apple Academic Press) and Ashish Kumar (Publisher, Apple Academic Press). To all of them sincere thanks and gratitude, as well as to all those who will find our scientific endeavor worthy of being continued and implemented for improving the environment life in general and of its accommodation with human social and economical interaction and evolution in special!

— Mihai V. Putz, PhD.

PART I
SPECTRAL-SAR ALGORITHM AND THE ALGEBRAIC CORRELATION FACTOR

Chapter 1

Introducing Spectral Structure-activity Relationship (SPECTRAL-SAR) and Algebraic Correlation Analysis: Connection with Computational Ecotoxicology

Mihai V. Putz and Ana-Maria Putz

INTRODUCTION

A novel quantitative structure-activity (property) relationship model, namely SPECTRAL-SAR (Special Computing Trace of Algebraic SAR), is presented in an exclusive algebraic way replacing the old-fashioned multi-regression one. The actual SPECTRAL-SAR method interprets structural descriptors as vectors in a generic data space that is further mapped into a full orthogonal space by means of the Gram-Schmidt algorithm. Then, by coordinated transformation between the data and orthogonal spaces, the SPECTRAL-SAR equation is given under simple determinant form for any chemical–biological interactions under study. While proving to give the same analytical equation and correlation results with standard multivariate statistics, the actual SPECTRAL-SAR frame allows the introduction of the spectral norm as a valid substitute for the correlation factor, while also having the advantage to design the various related SAR models through the introduced "minimal spectral path" rule. An application is given performing a complete SPECTRAL-SAR analysis upon the *Tetrahymena pyriformis* ciliate species employing its reported ecotoxicity activities among relevant classes of xenobiotics. By representing the spectral norm of the endpoint models against the concerned structural coordinates, the obtained SPECTRAL-SAR endpoints hierarchy scheme opens the perspective to further design the ecotoxicological test batteries with organisms from different species.

In chemistry, the first systematic correlations come from Lavoisier's law of conservation of mass and energy, followed by the Dalton conception of structural matter. Nevertheless, Mendeleyev was the first one to place the *structure-activity relationships* (SARs) in the center of chemistry with his vision of the periodic table [1]. However, with the advent of quantum theory, the relations among elements of periods and down groups of periodic table acquired in-depth quantitative meaning, by relating the elementary electronic structure with the manifested atomic reactivity through, for instance, basic electronegativity and chemical hardness indices [2, 3]. This way, it appears that every aspect of chemical reactivity can be seen as a certain manifestation

Putz M.V., Lacrămă A.M. "Introducing Spectral Structure Activity Relationship (S-SAR) Analysis. Application to Ecotoxicology", *International Journal of Molecular Sciences*, 8 (2007) 363-391.

of the structure-property pair that is quantified since the derivation of the associated equation [4].

Yet, the current problem of science is to organize the huge amount of experimental information in comprehensive equations with a predictive value. At this point, the quantitative structure-activity relationships (QSARs) methods seem to offer the best key for unifying the chemical and biological interaction into a single *in vivo-in vitro* content [5-10].

However, although the main purpose of QSARs studies is all about finding structural parameters that best correlate with the activity/property of the interactions observed, a multitude of methods of attaining this goal have appeared. They struggle to identify the most appropriate manner of quantifying the causes in such a way that they may be reflected in the measurement with maximal accuracy or minimal error. Phenomenologically, these methods can be conceptually grouped into "classical" [11-19], "3-dimensional" [20-30], "decisional" [31-42], and "orthogonal" ones [43-57], together represented as in Figure 1.1.

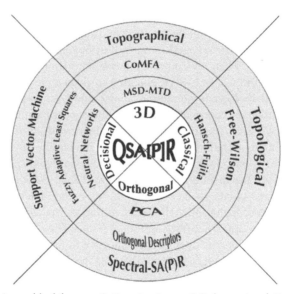

Figure 1.1. Generic world of the quantitative structure-activity/property relationships—QSA(P)R—through classical, 3D, decisional, and orthogonal methods of multivariate analysis of the chemical–biological interactions. In scheme MSD-MTD, CoMFA, and PCA stand for the "minimal steric difference-minimal topological difference," "comparative molecular field analysis," and "principal component analysis," respectively.

In short, classic QSAR approaches assume as descriptors the structural indices that directly reflect the electronic structures of the tested chemical compounds. As such, they assume that the biological activity depends on factors describing the lipophylicity (e.g., LogP, surfaces), electronic effects (e.g., Hammett constants, polarization, localization of charges), and steric effects (e.g., Taft indices, Verloop indices, topological indices, molecular mass, total energy at optimized molecular geometry) [12, 13].

A step forward is made when 3-dimensional structures are characterized by entry indices. For instance, the minimal topological difference (MTD) [23-25] and comparative molecular field analysis (CoMFA) [26, 27] methods are closely take into account the bioactive conformation of the receptor, the topology of the ligand series, as well as their steric fit, in accordance with the "key-into-lock" principle, while the topographical schemes [29, 30] make use of the graph representation of the chemical compounds, replacing in the associated connectivity matrices the optimized stereochemical indices. A visible increase in the structure-activity correlation is usually recorded when these methods are used. [28].

Still, statistically, it was found that in order for multiple linear regressions to be used, the requirement of a large number of compounds has to be met in order to explore the structural combination. Under these circumstances, the next QSARs category in Figure 1.1, namely the decisional one appears as further natural approach. Basically, they are heuristic methods of classifying data, developing genetic algorithms, that is neural networks [31, 32], fuzzy methodologies [34, 35], or support vector machine for learning [36-42], in order to find optimal solutions for combinatorial problems. They offer the advantage of providing a quick estimation regarding the quality of correlation we should expect from the data and furnishing several best regression models to decide upon. Moreover, the decisional analysis can be made in high-dimensional space always giving a solution by standard algorithm.

Nevertheless, despite having several solutions to decide over thousands of products from millions of libraries, together with hundred descriptors, that opens the problem of their further relevance and classification.

With these we have arrived at the heart of a QSAR analysis: the orthogonal problem. Statistically, this term was interpreted as descriptors whose values form a basis set that pose little inter-correlation factors. In practice, data reduction techniques such as principal component analysis (PCA) [43, 45] describe biological activity or chemical properties through a fewer number of independent (orthogonal) descriptors giving a regression equation on these principal components. Unfortunately, even combined with partial least squares (PLS) cross-validation technique to produce higher predictive QSAR models, the main drawback still remains since they furnish scarce possibility to interpret the obtained models [44, 46].

Another way of interpreting orthogonality was given through producing an orthogonal space by transforming the original basis set of descriptors in an orthogonal one by searching of inter-regression equations between them [47], followed, eventually, by their reciprocal subtractions [48].

Unfortunately, this method was found to give in almost all cases, the same correlation and statistical factors as those furnished by regressions with original basis set of descriptors [49–54], moreover, producing a QSAR equation in the orthogonal space where the orthogonal descriptors have little interpretation against the real ones. At the end, the orthogonal descriptors' method becomes another technique for selecting the independent predictor variables (like PCA) rather than one that provides alternative solution for basic SAR problem [55].

Under these circumstances, the third attempt of interpreting the orthogonal problem is considering the scalar product as the main vehicle in releasing the QSAR solution in a completely algebraic way thus furnishing the so called SPECTRAL-SAR technique in Figure 1.1 for reasons revealed bellow. It is based on the employment of the generalized Euclidian scalar product rule among the vectors associated to the descriptors' data in a way that produce, thought the Gram–Schmidt algorithm and coordinate transformation, precisely the same results as the statistical multi-linear regression techniques do. This new QSAR method, initiated in a relatively limited dissemination space [56, 57], is presented in full here, while also giving its equivalence with the standard multiple linear regression method.

Nevertheless, the features of the present SPECTRAL-SAR method include some of its predecessor's, including vectorial frame and output, high-dimensionality for the data space, adaptive analysis, showing, however, independence concerning the order of orthogonal vectors and also proving the spectral norm as an alternative algebraic tool for substituting of the statistical correlation factor.

The field of ecotoxicology was chosen as an application, where various combined SPECTRAL-SAR-Hansch models are constructed for describing the toxicity of 26 xenobiotics on the *Tetrahymena pyriformis* species. It follows that SPECTRAL-SAR approach gives the specific algebraic tool, that is the spectral norm, with which the specific ecotoxicological *endpoint* concept acquires new feasible degree.

The present SPECTRAL-SAR analysis leaves room for other similar studies when it is joined with other classical, 3-dimensional, and decisional QSAR techniques of Figure 1.1 so contributing to unite the chemical–biological interactions in a veritable QSAR science.

THE SPECTRAL-SAR METHOD

Background Concepts

The basic problem of SAR analysis can be formulated as follows: given a set of measured activities of a certain series of (say N) compounds, the optimal correlation between these activities and the structural (internal, intrinsic) properties of the compounds (say M properties) is sought, according to Table 1.1, in the form of the general multi-linear equation:

$$y = b_0 + b_1 x_1 + ... + b_k x_k + ... + b_M x_M + e . \tag{1.1}$$

In equation (1.1) y represents the generic activity in relation with an arbitrary set of independent variables x_i, i=1, ..., M through the fixed parameters b_j, j=0, ..., M, while e stands as the residual or error value between the assumed multi-linear model and measurements.

Therefore, the SAR problem becomes quantitative since the set of fixed parameters is determined so that the errors in activity evaluation are minimized. This way, the equation (1.1) may be used to predict the activity (without experimental measurement) for each further input of the structural parameters.

Table 1.1. Synopsis of the basic SAR descriptors.

Activity	Structural Predictor Variables				
y_1	x_{11}	$\cdots$	x_{1k}	$\cdots$	x_{1M}
y_2	x_{21}	$\cdots$	x_{2k}	$\cdots$	x_{2M}
$\vdots$	$\vdots$	$\vdots$	$\vdots$	$\vdots$	$\vdots$
y_N	x_{N1}	$\cdots$	x_{Nk}	$\cdots$	x_{NM}

However, this "Holy Grail" property of a QSAR equation opens the issue of significance and statistical relevance of the values considered in Table 1.1, as well as that of the computational method by which the parameters of (1.1) are assessed.

Usually, the QSAR problem is solved in the so called "normal" or "standard" way, briefly described in what follows. First, the equation (1.1) is particularized for each activity entry of Table 1.1 thus generating the $N \times (M+1)$ system:

$$
\begin{aligned}
y_1 &= b_0 + b_1 x_{11} + \ldots + b_k x_{1k} + \ldots + b_M x_{1M} + e_1 \\
y_2 &= b_0 + b_1 x_{21} + \ldots + b_k x_{2k} + \ldots + b_M x_{2M} + e_2 \\
y_N &= b_0 + b_1 x_{N1} + \ldots + b_k x_{Nk} + \ldots + b_M x_{NM} + e_N
\end{aligned}
\tag{1.2}
$$

Note that, generally, each activity evaluation is assumed to be accompanied by a different error, that is the values $e_1, \ldots, e_N$ are potentially different although the ideal case would demand that they be equal with zero.

However, since the following matrices are introduced

$$
Y = \begin{pmatrix} y_1 \\ y_2 \\ \vdots \\ y_N \end{pmatrix},\ E = \begin{pmatrix} e_1 \\ e_2 \\ \vdots \\ e_N \end{pmatrix},\ B = \begin{pmatrix} b_0 \\ b_1 \\ b_2 \\ \vdots \\ b_M \end{pmatrix},\ X = \begin{pmatrix} 1 & x_{11} & x_{12} & \cdots & x_{1M} \\ 1 & x_{21} & x_{22} & \cdots & x_{2M} \\ \vdots & \vdots & \vdots & \vdots & \vdots \\ 1 & x_{N1} & x_{N2} & \cdots & x_{NM} \end{pmatrix},
\tag{1.3}
$$

the system (1.2) can be rewritten in a simple algebraic way:

$$
Y = XB + E.
\tag{1.4}
$$

Hence, the minimization of the error vector E equals the minimization of the vector $(Y\text{-}XB)$ in (1.4). Put in vectorial terms, the solution of the supra-dimensional system (1.2) is a vector $\phi(B)$ which minimizes the Euclidian norm of the residual (error) vector, in a least square sense:

$$
\varphi(B) = (Y - XB)^T (Y - XB) \to \min.
\tag{1.5}
$$

Finally, one uses the following theorem [58–61]: *if the vector B of (1.3) is the solution of the linear system (1.6),*

$$X^T (Y - XB) = 0 \,, \qquad (1.6)$$

where X is a real matrix of dimension N × (M + 1) and B a vector of dimension (M + 1) × 1, then the standard deviation of XB with respect to Y is minimal, that is the condition (1.5) is fulfilled.

This means that we can consider $norm(E) \to 0$ when relations (1.4) and (1.6) are combined to give the B vector of estimates

$$B = \left(X^T X\right)^{-1} X^T Y \,. \qquad (1.7)$$

It is worth noting that while solution (1.7) solves the above QSAR problem in a formal way the concrete application of this method requires a high computational effort even when the symmetry of the matrix $X^T X$ is taken into account.

Despite this, the "normal" or "standard" QSAR procedure is already implemented in various software packages nowadays. It is worth exploring other alternative way that may serve both conceptual and computational advantages. The so called "spectral" algorithm, presented below, stands as such a new perspective, belonging to "orthogonal QSAR" methods of Figure 1.1.

SPECTRAL-SAR Algorithm

The key concept in SAR discussion regards the independence of the considered structural parameters in Table 1.1. As a consequence, we may further employ this feature to quantify the basic SAR through an orthogonal space.

The idea is to transform the columns of structural data of Table 1.1 into an abstract orthogonal space, where necessarily all predictor variables are independent, see Figure 1.2; solve the SAR problem there and then referring the result to the initial data by means of a coordinate transformation.

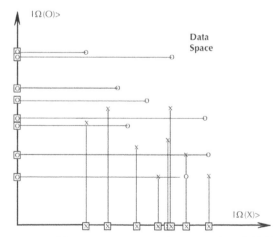

Figure 1.2. Generic mapping of data space containing the vectorial sets $\{|X\rangle, |0\rangle\}$ into orthogonal basis $\{|\Omega(X)\rangle, |\Omega(O)\rangle\}$.

The analytical procedure is unfolded in simple tree steps.

Basically, Table 1.1 is reconsidered under the form of Table 1.2 where, for completeness, the unity column has been added $|X_0\rangle = |1 \quad 1 \quad ... \quad 1\rangle$ for accounting of the coefficients of the free term (b_0) of system (1.2).

Table 1.2. The spectral (vectorial) version of SAR descriptors of Table 1.1.

Activity	Structural Predictor Variables					
$|Y\rangle$	$|X_0\rangle$	$|X_1\rangle$	...	$|X_k\rangle$	...	$|X_M\rangle$
y_1	1	x_{11}	...	x_{1k}	...	x_{1M}
y_2	1	x_{21}	...	x_{2k}	...	x_{2M}
$\vdots$	$\vdots$	$\vdots$	$\vdots$	$\vdots$	$\vdots$	$\vdots$
y_N	1	x_{N1}	...	x_{Nk}	...	x_{NM}

Moreover, since the columns are now considered as vectors in data space we are looking for the "spectral" decomposition of the activity vector $|Y\rangle$ upon the considered basis of the structural vectors $\{|X_0\rangle,|X_1\rangle,...,|X_k\rangle,...,|X_M\rangle\}$:

$$|Y\rangle = b_0|X_0\rangle + b_1|X_1\rangle + ... + b_k|X_k\rangle + ... + b_M|X_M\rangle + |e\rangle. \qquad (1.8)$$

Equation (1.8) stands, in fact, as a spectral decomposition counterpart of the multilinear equation (1.1), equation that the name of the present approach comes from.

The next step is to construct a vectorial algorithm so that the residual vector $|e\rangle$ can be sent to zero in (1.8) in order to fulfill the above (1.5) condition of minimizing of errors.

To achieve the minimal errors in (1.8) the transformation of the data basis $\{|X_0\rangle,|X_1\rangle,...,|X_k\rangle,...,|X_M\rangle\}$ into an orthogonal one, say $\{|\Omega_0\rangle,|\Omega_1\rangle,...,|\Omega_k\rangle,...,|\Omega_M\rangle\}$, is now considered. In this respect the consecrated Gram-Schmidt procedure is employed. It is worth noting that this procedure is well known in quantum chemistry when searching for an orthogonal basis for an orthogonal basis set in atomic and molecular wave function spectral decomposition [62].

However, before applying it effectively one has to introduce the generalized scalar product throughout the basic rule:

$$\langle \Psi_l | \Psi_k \rangle = \sum_{i=1}^{N} \psi_{il}\psi_{ik} = \langle \Psi_k | \Psi_l \rangle \qquad (1.9)$$

giving out a real number from two arbitrary N-dimensional vectors

$$|\Psi_l\rangle = |\psi_{1l} \quad \psi_{2l} \quad ... \quad \psi_{Nl}\rangle |\Psi_k\rangle = |\psi_{1k} \quad \psi_{2k} \quad ... \quad \psi_{Nk}\rangle.$$

Briefly, remember that the orthogonal condition requires that the scalar product of type (1.9) to be zero, the orthogonal basis $\{|\Omega_0\rangle, |\Omega_1\rangle, ..., |\Omega_k\rangle, ..., |\Omega_M\rangle\}$ can be constructed from the set $\{|X_0\rangle, |X_1\rangle, ..., |X_k\rangle, ..., |X_M\rangle\}$ according with the iterative recipe:

i. Choose

$$|\Omega_0\rangle = |X_0\rangle;$$
(1.10)

ii. Then, by picking $|X_1\rangle$ as the next vector to be transformed, one can write that:

$$|\Omega_1\rangle = |X_1\rangle - r_0^1 |\Omega_0\rangle, \; r_0^1 = \frac{\langle X_1|\Omega_0\rangle}{\langle \Omega_0|\Omega_0\rangle}$$
(1.11)

so that $\langle \Omega_0|\Omega_1\rangle = 0$ assuring so far that $|\Omega_0\rangle$ and $|\Omega_1\rangle$ are orthogonal.

iii. Next, repeating steps i. and ii. above until the vectors $|\Omega_0\rangle$, $|\Omega_1\rangle$, ..., $|\Omega_{k-1}\rangle$ are orthogonally constructed, we can, for instance, further transform the vector $|X_k\rangle$ into:

$$|\Omega_k\rangle = |X_k\rangle - \sum_{i=0}^{k-1} r_i^k |\Omega_i\rangle, \; r_0^1 = \frac{\langle X_1|\Omega_0\rangle}{\langle \Omega_0|\Omega_0\rangle}$$
(1.12)

so that the vector $|\Omega_k\rangle$ is orthogonal on all previous ones.

iv. Step (iii) is repeated and extended until the last orthogonal predictor vector $|\Omega_M\rangle$ is obtained.

Therefore, grounded on the Gram-Schmidt recipe the starting predictor vectorial basis $\{|X_0\rangle, |X_1\rangle, ..., |X_k\rangle, ..., |X_M\rangle\}$ is replaced with the orthogonal one $\{|\Omega_0\rangle, |\Omega_1\rangle, ..., |\Omega_k\rangle, ..., |\Omega_M\rangle\}$ by appropriately subtracting from the original vectors the non-wished non-orthogonal contributions. Note that the above procedure holds for any arbitrary order of original vectors to be orthogonalized.

Within the constructed orthogonal space, the vector activity $|Y\rangle$ achieves true spectral decomposition form:

$$|Y\rangle = \omega_0 |\Omega_0\rangle + \omega_1 |\Omega_1\rangle + ... + \omega_k |\Omega_k\rangle + ... + \omega_M |\Omega_M\rangle.$$
(1.13)

Note that the residual vector in equation (1.8) has disappeared in (1.13) since it has no structural meaning in the abstracted orthogonal basis. Or, alternatively, one can say that in the abstract orthogonal space the residual vector $|e\rangle$ was identified with the vector with all components zero $|0,0,...,0\rangle$ that is always perpendicular with all other vectors of orthogonal basis.

This way, the Gram-Schmidt algorithm, by its specific orthogonal recursive rules, absorbs, or transforms the minimization condition of errors in (1.8) to simple identification with the origin of the orthogonal space of data.

At this point, since there is no residual vector remaining in (1.13) one can consider that the SAR problem is in principle solved once the new coefficients in (1.13) ($\omega_0, \omega_1, ..., \omega_k, ..., \omega_M$) are determined. These new coefficients can be immediately deduced based on the orthogonal peculiarities of the spectral decomposition (1.13) grounded on the fact that:

$$\langle \Omega_k | \Omega_l \rangle = 0 \, , k \neq l \, , \tag{1.14}$$

a condition assured by the very nature of the vectors from the constructed orthogonal basis.

As such, each coefficient comes out as the scalar product of its specific predictor vector with the activity vector (1.13) is performed:

$$\omega_k = \frac{\langle \Omega_k | Y \rangle}{\langle \Omega_k | \Omega_k \rangle}, k = \overline{0, M} \, . \tag{1.15}$$

With coefficients given by expressions of type (1.15) the spectral expansion of the activity vector into an orthogonal basis (1.13) is completed. Yet, this does not mean that we have found the coefficients that directly link the activity with the predictor vectors as equation (1.8) demands.

However, this goal is easily achieved through the final stage of the present SAR algorithm. It consists in going back from the orthogonal to the initial basis of data through the system of coordinate transformations:

$$\begin{cases} |Y\rangle = \omega_0 |\Omega_0\rangle + \omega_1 |\Omega_1\rangle + ... + \omega_k |\Omega_k\rangle + ... + \omega_M |\Omega_M\rangle \\ |X_0\rangle = 1 \times |\Omega_0\rangle + 0 \times |\Omega_1\rangle + ... + 0 \times |\Omega_k\rangle + ... + 0 \times |\Omega_M\rangle \\ |X_1\rangle = r_0^1 |\Omega_0\rangle + 1 \times |\Omega_1\rangle + ... + 0 \times |\Omega_k\rangle + ... + 0 \times |\Omega_M\rangle \\ \quad .... \\ |X_k\rangle = r_0^k |\Omega_0\rangle + r_1^k |\Omega_1\rangle + ... + 1 \times |\Omega_k\rangle + ... + 0 \times |\Omega_M\rangle \\ \quad .... \\ |X_M\rangle = r_0^M |\Omega_0\rangle + r_1^M |\Omega_1\rangle + ... + r_k^M |\Omega_k\rangle + ... + 1 \times |\Omega_M\rangle \end{cases} \tag{1.16}$$

While the first equation of (1.16) reproduces the entire spectral decomposition (1.13), the rest of them are convenient rewritings of the Gram–Schmidt transformations (1.10)–(1.12).

Finally, the system (1.16) is algebraically true if and only if the associated augmented determinant disappears,

$$
\begin{vmatrix}
|Y\rangle & \omega_0 & \omega_1 & \cdots & \omega_k & \cdots & \omega_M \\
|X_0\rangle & 1 & 0 & \cdots & 0 & \cdots & 0 \\
|X_1\rangle & r_0^1 & 1 & \cdots & 0 & \cdots & 0 \\
\vdots & \vdots & \vdots & \vdots & & \vdots & \\
|X_k\rangle & r_0^k & r_1^k & \cdots & 1 & \cdots & 0 \\
\vdots & \vdots & \vdots & \vdots & & \vdots & \\
|X_M\rangle & r_0^M & r_1^M & \cdots & r_k^M & \cdots & 1
\end{vmatrix} = 0 ,
\tag{1.17}
$$

this being the condition consecrated by the theorem according to which *a system possessing a column as a linear combination of all others has a zero allied determinant* [58].

It is worth noting that the minimization of residual errors was unnecessarily complicated in previous orthogonalization approaches [54] by involving standard multilinear regressions, iteratively, among the selected structural descriptors and of their combination [53]. This way, the flavor of performing an alternative orthogonal approach of the SAR issue was lost in an ocean of inter-correlations. Consequently, the heuristic methods used in the search for an orthogonal set of descriptors, in the regression sense, though an arbitrarily minimal inter-correlation factors, leave both the realistic meaning of the usually lesser set of orthogonalized descriptors, as compared with the original one, and the initial SAR problem to be solved. On the contrary, within the present orthogonal endeavor, the SPECTRAL-SAR method proposes a new way of completely solving of a SAR problem linking the measured activities (or observed properties) with the structural descriptors in a simpler and more transparent algebraic way than the "standard" multi-linear regression method do.

Moreover, the ordering problem in all previous orthogonal descriptors' methods [54] is eliminated with the present SPECTRAL-SAR analysis since all structural descriptors are spectrally expanded at once complying with the orthogonal basis, as equation (1.16) reveals, avoiding iterative reciprocal correlations among orthogonal descriptors where their considered order becomes essential. This special feature of SPECTRAL-SAR will be illustrated later, in the application section.

It is now clear that once expanded, observing its first column, the determinant (1.17) generates the searched full solution of the basic SAR problem of Table 1.2 with minimization of errors included and independent of the orthogonalization order. Remarkably, apart from being conceptually new through considering the spectral (orthogonal) expansion of the input data space (of both activity and descriptors) through the system (1.16), the present method also has the computational advantage of being simpler than the classical "standard" way of treating SAR problem previously exposed. That because, one has nothing to do with computations of matrix of the coefficients (1.7), this being a quite involving and time consuming procedure. Instead, one can write directly the SPECTRAL-SAR solution (equation) as the expansion of a $(M + 2)$-dimensional determinant of type (1.17) whose components are the activity and structural vectors among the Gram–Schmidt and the spectral decomposition coefficients, r_k^l and ω_k, respectively.

However, although different from the mathematical procedure, both standard- and SPECTRAL-SAR give similar results due to the theorem that states that [61]: *if the matrix X, as that from (1.3), with dimension $N \times (M + 1)$, $N > M + 1$, has linear independent columns*, that is they are orthogonal as in the spectral approach, *then there exists an unique matrix Q of dimension $N \times (M + 1)$ with orthogonal columns and a triangular matrix R of dimension $(M + 1) \times (M + 1)$ with the elements of the principal diagonal equal with 1*, as identified in the first small determinant in (1.17), *so that the matrix X can be factorized as*

$$X = QR. \tag{1.18}$$

When combining equation (1.18) with the optimal equation (1.6) one can get, after straight algebraic rules, that the B vector of estimates takes the form

$$B = \left(Q^T Q\right)^{-1} Q^T Y, \tag{1.19}$$

in close agreement with previous normal one, see equation (1.7). However, by comparison of matrices $X^T X$ and $Q^T Q$ in equations (1.7) and (1.19), respectively, there is clear that the last case certainly furnishes a diagonal form which for sure is easier to handle (i.e., to take its inverse) when searching for the vector B of SAR coefficients.

With these considerations one would prefer the present SPECTRAL-SAR approach when solving the QSAR problems in chemistry and related molecular fields. Nevertheless, wishing to also provide a practical advantage of the exposed SPECTRAL-SAR scheme, a specific application, with relevance in ecotoxicological studies, is presented in the next section.

APPLICATION TO ECOTOXICOLOGY

Basic Characteristics of QSAR in Ecotoxicology

From more than one decade the European Union institutions, for example Organization for Economic Co-operation and Development (OECD) through its Registration, Evaluation, and Authorization of Chemicals (REACH) management system [63, 64], the United States Environmental Protection Agency (EPA) as part of the premanufactory notification assessment, as well as the World Health Organization have been developing impressive programs on the regulatory assessment of chemical safety by using of the QSAR data bases and of the associated automated expert systems [65–73]. This because, with the tones of chemicals that force their way onto the market each year and due to their commercial and industrial disposal into the environment, it becomes of first importance to predict their toxicological activities from the molecular structure in order to properly design the risk assessment measures [67–77].

Nevertheless, in order to best accomplish such a goal, both a conceptual and a computational strategy need to be adopted. As such, while, for instance, a certain set of parameters has been identified for environmental studies, that is bioaccumulation, chemical degradation (aqueous and gas phase), biodegradation, soil sorption, and ecotoxicity, two major aspects have been identified for QSAR analyses, namely the quality and the chemical domain of the QSAR [69, 71, 72].

Concerning the parameters to be evaluated, they are analytically transposed into the so called *endpoints*, representing specific experimental and measurement quantities giving information about the environmental risk degree. They are thus identified with the QSAR activities (biophores or toxicophores) to be correlated and are usually expressed as log-based continuous toxicological data (e.g., median lethal concentration-LC_{50}, 50% effect concentration-EC_{50}, 50% grow inhibition concentration-IGC_{50}) [74-77].

On the other hand, a useful QSAR model has to satisfy selection criteria in order to be validated.

From the statistical point of view the ratio of data points to the number of variables should be higher or equal to 5 (the so called Topliss–Costello rule [78]) and to provide a correlation factor $r > 0.84$.

As descriptors, those directly related to molecular structure of chemical are preferable. It is worth noting here that the quantum chemical parameters have an advantage against those of topological nature; still the quantum parameters to be used has to be relatively easily obtainable, for instance those based on ground state or valence state properties of compounds are preferable to those based on transition-state calculations [10].

If descriptors are taken from experiments, the experimental conditions must be specified. Nevertheless, the best models predicting ecotoxic effects have to be mechanistic interpretable, though that structure-activity correlation permits reconstruction or prediction of the basic phenomena that take place at the molecular level.

Regarding the outliers they have to be treated with caution, as they are not necessarily outside of the chemical domain but depending on the QSAR model (i.e., of the correlated descriptors) employed [79]. Moreover, the atypical data (presumed outliers) may represent compounds acting by a different mechanism, inducing an inhibition or belonging to dissimilar chemical structure. However, they should not be excluded from an analysis unless relevant alternative QSAR models were constructed. With this issue, we arrive at the chemical domain problem or at the representative set of compounds for the QSAR analysis.

Based on previous criteria in order for a QSAR analysis to be well conducted, a compromise between breath (variety) and depth (representability) characteristics through the existing chemicals within that domain have to be considered.

This way, the two-fold process of dissimilarity- and similarity-based selection is achieved [10]. The motivation for this criteria is that, while similar compounds (usually based on substitutions) assures the basic congenericity QSAR condition, considering dissimilar chemicals can predict how (however subtle) alterations in molecular structure can lead to changes in the mechanism of toxicity action and potency in the tested series of compounds. In short, this condition can be regarded as structural heterogeneity of compounds.

After all, it is widely recognized that ecotoxicity action is a multivariate process involving xenobiotics leading with immediate and long-term effects due to various transformations products. Therefore, a QSAR approach may provide information of the bio-up-take (i.e., of key process) through the selected descriptors that can be integrated in an expert system of toxic prediction.

However, with a view to designing an ecotoxicological mechanistic battery for different species on QSAR grounds, the first stage of unicellular organism level is undertaken here.

Bio-ecological Issues of Unicellular Organisms

We often think of unicellular organisms as having a simple, primitive structure. This is definitely an erroneous view when applied to the ciliates; they are probably the most complex of all unicellular organisms.

Unlike multicellular organisms, which have cells specialized for performing the various body functions, single-celled organisms must perform all these functions with a single cell, and so their structure may be much more complex than the cells of larger organisms.

Movement, sensitivity to the environment, water balance, and food capture must all be accomplished with the machinery in a single cell [80, 81a–d]. As protozoans, these organisms are classified according to their means of locomotion: by cilia (*Ciliophora*), flagella (*Sarcomastigophora*), or pseudopodia (*Rhizopoda*), while non-motile protists are classified as sporozoans in the phylum *Apicomplexa*.

Many of these single-celled organisms feed by engulfing smaller organisms directly into temporary intracellular vacuoles. These food vacuoles circulate in a characteristic manner within the cells while enzymes are secreted into them for digestion [81b].

However, form the taxonomy points of view they are classified downwards, from kingdom to species as: *Protista > Ciliophora > Cyrtophora > Oligohymenophorea > Hymenostomatia > Hymenostomatida > Tetrahymeni > Tetrahymenidae > Tetrahymena* [81c].

However, it is worth restricting the discussion to ciliates only since they include about 7500 known species of some of the most complex single-celled organisms ever, as well as some of the largest free-living protists; a few genera may reach 2 mL in length, and are abundant in almost every environment with liquid water: ocean waters, marine sediments, lakes, ponds, and rivers, and even soils. Because individual ciliate species vary greatly in their tolerance of pollution, the ciliates found in a body of water can be used to gauge the degree of pollution quickly.

More specifically, ciliates are classified on the basis of cilia arrangement, position, and ultrastructure. Such work now involves electron microscopy and comparative molecular biology to estimate relationships.

In the most recent classification of ciliates, the group is divided into eight classes: *Prostomatea Benthic* and *Karyorelictida Benthic* (mostly in marine forms), *Litostomatea* (including *Balantidium* and *Didinium*), *Spirotrichea* (including *Stentor*, *Stylonychia*, and tintinnids), *Phyllopharyngea* (including suctorians), *Nassophorea* (including *Paramecium* and *Euplotes*), *Oligohymenophorea* (including *Tetrahymena*, *Vorticella*, and *Colpidium*), and Colpodea (including *Colpoda*) [81a].

Nevertheless, most frequently studied unicellular organisms through QSAR toxicological analysis are from the *Tetrahymena* genus of ciliated protozoa. All species of the genus *Tetrahymena* are morphologically very similar; they display multiple nuclei: a diploid micronucleus found only in conjugating strains and a polyploid macronucleus present in all strains, which is the site of gene expression during vegetative growth, see Figure 1.3 [82, 83].

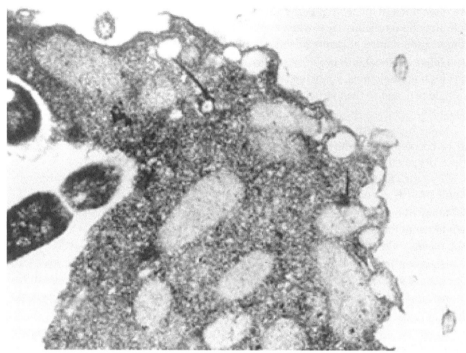

Figure 1.3. Illustration of the oral region of *Tetrahymena pyriformis* during ingestion as taken by electron micrograph technique [83].

Tetrahymena species are very common in aquatic habitats and are non-pathogenic, have a short generation time and can be grown to high cell density in inexpensive media [81d]. As such, ecological, morphological, biochemical, and molecular features have been used over the years in attempts to classify them.

The earliest classifications were based on morphological and ecological data. At this level the presence or absence of a caudal cilium was regarded as an important character. Later, three morphological species complexes were distinguished: the pyriformis complex with smaller, bacterivorous species and less somatic kinetics; the rostrata complex with larger parasitic or histophagous species, more somatic kinetics, and the ability to form resting cysts; and the patula complex with species that undergo microstome–macrostome transformation. Within the complexes, particularly the pyriformis complex, species are distinguishable by their mating capacity and/or isozyme patterns. Finally, another approach based on the degree of parasitism was suggested. Since, the *Tetrahymena* species are free-living, as well as facultative and obligate parasites, it was suggested an evolutionary lineage from free-living species, considering *Tetrahymena pyriformis* to be the basal species, to facultative parasites, and then to obligate parasites [80, 81a, 82, 84]. Accordingly, *Tetrahymena pyriformis*, a teardrop-shaped, unicellular, ciliated freshwater protozoan about 50 μm long, is found as the best candidate whose ecotoxicological activity is considered through the present SPECTRAL-SAR toward establishing a mechanistically coherent view of a certain class of xenobiotics on inter-correlated species.

SPECTRAL-SAR Ecotoxicity of *Tetrahymena pyriformis*

Quite often, despite the tendency to submit a large class of descriptors to a QSAR analysis, this is not the best strategy [69], at least in ecotoxicology, and whenever a specific mode of action or the elucidation of the causal mechanistically scheme is envisaged.

More focused studies in ecotoxicology, and especially regarding *Tetrahymena pyriformis*, have found that hydrophobicity (*LogP*) and electrophilicity (E_{LUMO}) phenomena plays a particular place in explaining the ecotoxicology of the species.

While hydrophobicity describes the penetration power of the xenobiotics though biological membranes, the other descriptors to be considered reflect the electronic and specific interaction between the ligand and target site of receptor.

Moreover, it was convincingly argued that the classical Hammett constant can be successfully rationalized by a pure structural index as the energy of the lowest unoccupied molecular orbital (*LUMO*) is [79]. These facts open the attractive perspective of considering the ecotoxicological studies through employing the Hansch-type structure-activity expansion:

$$A = b_0 + b_1 \left(\frac{hydrophobic}{descriptor} \right) + b_2 \left(\frac{electronic}{descriptor} \right) + b_3 \left(\frac{steric}{descriptor} \right), \tag{1.20}$$

thus also providing enough information from transport, electronic affinity, and specific interaction at the molecular level, respectively.

However, in the present study, besides considering *LogP* as compulsory descriptor the molecular polarizability (POL) will be considered for modeling the electronic affinity for its inherent definition that implies the radius of the electrostatic sphere of electrostatic interaction. This way, the first stage of binding, through the radius of interaction, is accounted [85].

Then, the steric descriptor is chosen here, for simplicity, as the total molecular energy (E_{TOT}) in its ground state, for the reason that it is calculated at the optimum molecular geometry where the stereo-specificity is included.

Under these circumstances the ecotoxic activity to *Tetrahymena pyriformis*, determined in a population growth impairment assay with a 40 h static design and population density measured spectrophotometrically as the endpoint $A = \text{Log} (1/IGC_{50})$ [86–90], from a series of xenobiotics of which majority are of phenol type is in Table 1.3 considered.

Table 1.3. The series of the xenobiotics of those toxic activities $A = Log(1/IGC_{50})$ were considered [86] along structural parameters $LogP$, POL (E^3), and E_{TOT} (kcal/mol) as accounting for the hydrophobicity, electronic (polarizability), and steric (total energy at optimized 3D geometry) effects, respectively, derived with the help of HyperChem program [91].

No.	Compound		A	$\lvert 1 \rangle$	$Log P$	POL	E_{TOT}
	Name	*Formulae*	$\lvert Y \rangle$	$\lvert X_1 \rangle$	$\lvert X_1 \rangle$	$\lvert X_2 \rangle$	$\lvert X_3 \rangle$
1	methanol	CH_3OH	−2.67	1	−0.27	3.25	−11622.9
2	ethanol	C_2H_5OH	−1.99	1	0.08	5.08	−15215.4
3	butan-1-ol	C_4H_9OH	−1.43	1	0.94	8.75	−22402.8
4	butanone	C_4H_8O	−1.75	1	1.01	8.2	−21751.8
5	pentan-3-one	$C_5H_{10}O$	−1.46	1	1.64	10.04	−25344.6
6	phenol	C_6H_5OH	−0.21	1	1.76	11.07	−27003.1
7	aniline	$C_6H_5NH_2$	−0.23	1	1.26	11.79	−24705.9
8	3-cresol	$CH_3\text{-}C_6H_4\text{-}OH$	−0.06	1	2.23	12.91	−30597.6
9	4-methoxiphenol	$OH\text{-}C_6H_4\text{-}O\text{-}CH_3$	−0.14	1	1.51	13.54	−37976.3
10	2-hydroxyaniline	$OH\text{-}C_6H_4\text{-}NH_2$	0.94	1	0.98	12.42	−32095.4
11	Benzaldehyde	$C_6H_5\text{-}CHO$	−0.2	1	1.72	12.36	−29946.9
12	2-cresol	$CH_3\text{-}C_6H_4\text{-}OH$	−0.27	1	2.23	12.91	−30597.2
13	3,4-dimeyhylphenol	$C_6H_3(CH_3)_2OH$	0.12	1	2.7	14.74	−34190.8
14	3-nitrotoluene	$CH_3\text{-}C_6H_4\text{-}NO_2$	0.05	1	0.94	13.98	−42365.1
15	4-chlorophenol	$C_6H_5\text{-}O\text{-}Cl$	0.55	1	2.28	13	−35307.6
16	2,4-dinitroaniline	$C_6H_3(NO_2)NH_2$	0.53	1	−1.75	15.22	−63030.2
17	2-methyl-1-4-naphtoqui-none	$C_{11}H_8O_2$	1.54	1	2.39	20.99	−49768.3
18	1,2-dichlorobenzene	$C_6H_4Cl_2$	0.53	1	3.08	14.29	−36217.2
19	2,4-dinitrophenol	$C_6H_3(NO_2)OH$	1.08	1	1.67	14.5	−65318
20	1,4-dinitrobenzene	$C_6H_4N_2O_4$	1.3	1	1.95	13.86	−57926.7
21	2,4-dinitrotoluene	$C_7H_6(NO_2)_2$	0.87	1	2.42	15.7	−61520.7
22	2,6-ditertbutil 4-methyl phenol	$C_{15}H_{23}OH$	1.8	1	5.48	27.59	−59316.5
23	2,3,5,6-tetrachloroaniline	$C_6H_3NCl_4$	1.76	1	3.34	19.5	−57920.2
24	penthachlorophenol	C_6Cl_5OH	2.05	1	−0.54	20.71	−68512.4
25	phenylazophenol	$C_{12}H_{10}N_2O$	1.66	1	4.06	22.79	−55488.9
26	pentabromophenol	C_6Br_5OH	2.66	1	5.72	24.2	−66151.5

It is worth mentioning that the number of compounds is in relevant ratio with the number of descriptors used, according with above Topliss–Costello rule, and that both chemical variability and congenericity are fulfilled since most of them reflect the phenolic toxicity.

The standard QSAR analysis of data of Table 1.3 for all possible models of actions reveals the multivariate equations displayed in Table 1.4, together with their associate statistics:

$$r = \sqrt{1 - \frac{SR}{SQ}} ,$$

(1.21)

$$s = \sqrt{\frac{SR}{N - M - 1}}$$

(1.22)

$$F_{M, N-M-1} = \frac{N - M - 1}{M} \left(\frac{SQ}{SR} - 1 \right)$$

(1.23)

as correlation factor, standard error of estimate, and Fisher index, respectively, in terms of the total number of residues, measuring the spreading of the input activities with respect to their estimated counterparts,

$$SR = \sum_{i=1}^{N} \left(A_i - A_i^{PREDICTED} \right)^2$$

(1.24)

and the total sum of squares,

$$SQ = \sum_{i=1}^{N} \left(A_i - \overline{A} \right)^2 ,$$

(1.25)

measuring the dispersion of the measured activities around their average:

$$\overline{A} = \frac{1}{N} \sum_{i=1}^{N} A_i .$$

(1.26)

while the number of compounds and descriptors were fixed to $N = 26$ and $M = 3$, in each endpoint case, respectively.

Table 1.4. QSAR equations through standard multi-linear routine of Satistical package [92] for all possible correlation models considered from data of Table 1.3.

Model	Variables	QSAR Equation	r	s	F
Ia	logP	$A^{Ia} = -0.547836 + 0.435669logP$	0.539	1.15	9.834
Ib	POL	$A^{Ib} = -2.84021 + 0.2166POL$	0.908	0.574	112.15
Ic	E_{TOT}	$A^{Ic} = -2.50233 - 0.00007\, E_{TOT}$	0.882	0.644	84.015
IIa	logP, POL	$A^{IIa} = -2.91377 - 0.08109logP$ $+0.23233POL$	0.911	0.58	55.930

Table 1.4. *(Continued)*

Model	Variables	QSAR Equation	r	s	F
IIb	logP, E_{TOT}	$A^{IIb} = -2.64602 + 0.22991 logP$ $-0.00006\, E_{TOT}$	0.922	0.54	65.339
IIc	POL, E_{TOT}	$A^{IIc} = -2.98407 + 0.13427 POL$ $-0.00003\, E_{TOT}$	0.939	0.478	86.503
III	logP, POL, E_{TOT}	$A^{III} = -2.94395 + 0.06335 logP$ $+0.11206 POL$ $-0.00004 E_{TOT}$	0.941	0.48	56.598

Before attempting a mechanistic analysis of the results, let us apply the SPECTRAL-SAR techniques to the same data of Table 1.3 by using the key (or spectral) equation-type (1.17) with the associated determinant completed with orthogonal and spectral coefficients of equations (1.12) and (1.15), in each considered model of ecotoxic action, respectively.

More explicitly, in equations (1.27)–(1.29), the spectral equations are presented with their determinant forms that once expanded produce the spectral multi-linear dependencies of Table 1.5.

(Ia):

$$\begin{vmatrix} \left|Y\right\rangle^{Ia} & 0.270385 & 0.435669 \\ \left|X_0\right\rangle & 1 & 0 \\ \left|X_1\right\rangle & 1.87808 & 1 \end{vmatrix} = 0 \text{ or } \begin{vmatrix} \left|Y\right\rangle^{Ia} & 0.268751 & -0.547836 \\ \left|X_1\right\rangle & 1 & 0 \\ \left|X_0\right\rangle & 0.304687 & 1 \end{vmatrix} = 0, \quad (27a)$$

(Ib):

$$\begin{vmatrix} \left|Y\right\rangle^{Ib} & 0.270385 & 0.216598 \\ \left|X_0\right\rangle & 1 & 0 \\ \left|X_2\right\rangle & 14.3612 & 1 \end{vmatrix} = 0 \text{ or } \begin{vmatrix} \left|Y\right\rangle^{Ib} & 0.0441181 & -2.84021 \\ \left|X_2\right\rangle & 1 & 0 \\ \left|X_0\right\rangle & 0.0607278 & 1 \end{vmatrix} = 0, \quad (1.27b)$$

(Ic):

$$\begin{vmatrix} \left|Y\right\rangle^{Ic} & 0.270385 & -0.000067863 \\ \left|X_0\right\rangle & 1 & 0 \\ \left|X_3\right\rangle & -40857.5 & 1 \end{vmatrix} = 0 \text{ or } \begin{vmatrix} \left|Y\right\rangle^{Ic} & -0.0000157064 & -2.50233 \\ \left|X_3\right\rangle & 1 & 0 \\ \left|X_0\right\rangle & -0.0000208433 & 1 \end{vmatrix} = 0 \quad (1.27c)$$

(IIa):

$$
\begin{vmatrix}
|Y\rangle^{IIa} & 0.270385 & 0.435669 & 0.232325 \\
|X_0\rangle & 1 & 0 & 0 \\
|X_1\rangle & 1.87808 & 1 & 0 \\
|X_2\rangle & 14.3612 & 2.22431 & 1
\end{vmatrix} = 0 ,
\qquad (1.28a)
$$

(IIb):

$$
\begin{vmatrix}
|Y\rangle^{IIb} & 0.270385 & 0.435669 & -0.0000608117 \\
|X_0\rangle & 1 & 0 & 0 \\
|X_1\rangle & 1.87808 & 1 & 0 \\
|X_3\rangle & -40857.5 & -3383.5 & 1
\end{vmatrix} = 0 ,
\qquad (1.28b)
$$

(IIc):

$$
\begin{vmatrix}
|Y\rangle^{IIc} & 0.270385 & 0.216598 & -0.0000324573 \\
|X_0\rangle & 1 & 0 & 0 \\
|X_2\rangle & 14.3612 & 1 & 0 \\
|X_3\rangle & -40857.5 & -2536.37 & 1
\end{vmatrix} = 0 ,
\qquad (1.28c)
$$

(III):

$$
\begin{vmatrix}
|Y\rangle^{III} & 0.270385 & 0.435669 & 0.232325 & -0.0000363728 \\
|X_0\rangle & 1 & 0 & 0 & 0 \\
|X_1\rangle & 1.87808 & 1 & 0 & 0 \\
|X_2\rangle & 14.3612 & 2.22431 & 1 & 0 \\
|X_3\rangle & -40857.5 & -3383.5 & -3306.57 & 1
\end{vmatrix} = 0 .
\qquad (1.29)
$$

Remarkably, one may easily note the striking similitude of the equations in Tables 1.4 and 1.5, respectively. Moreover, in equations (1.27) the spectral determinant was written in both possible ways of orthogonalization, nevertheless leading to the same results in Table 1.5. That is the computational proof that SPECTRAL-SAR indeed provides a viable alternative to standard QSAR at each level of modeling, being independent of number of descriptors, compounds, or order of orthogonalization. We advocate on the computational advantage of SPECTRAL-SAR though lesser steps of computation and by the full analyticity of the delivered structure-activity equation, through a simple transparent determinant.

Table 1.5. Spectral structure-activity relationships (SPECTRAL-SAR) through determinants of Equations (1.27)–(1.29) for all possible correlation models considered from the data in Table 1.3.

Models	Vectors	SPECTRAL-SAR Equation
Ia	$\left\lvert X_0\right\rangle, \left\lvert X_1\right\rangle$	$\left\lvert Y\right\rangle^{Ia} = -0.547836\left\lvert X_0\right\rangle + 0.435669\left\lvert X_1\right\rangle$
Ib	$\left\lvert X_0\right\rangle, \left\lvert X_2\right\rangle$	$\left\lvert Y\right\rangle^{Ib} = -2.84021\left\lvert X_0\right\rangle + 0.216598\left\lvert X_2\right\rangle$
Ic	$\left\lvert X_0\right\rangle, \left\lvert X_3\right\rangle$	$\left\lvert Y\right\rangle^{Ic} = -2.50233\left\lvert X_0\right\rangle - 0.000067863\left\lvert X_3\right\rangle$
IIa	$\left\lvert X_0\right\rangle, \left\lvert X_1\right\rangle, \left\lvert X_2\right\rangle$	$\left\lvert Y\right\rangle^{IIa} = -2.91377\left\lvert X_0\right\rangle - 0.0810929\left\lvert X_2\right\rangle + 0.232325\left\lvert X_2\right\rangle$
IIb	$\left\lvert X_0\right\rangle, \left\lvert X_1\right\rangle, \left\lvert X_3\right\rangle$	$\left\lvert Y\right\rangle^{IIb} = -2.64602\left\lvert X_0\right\rangle + 0.229913\left\lvert X_1\right\rangle - 0.0000608117\left\lvert X_3\right\rangle$
IIc	$\left\lvert X_0\right\rangle, \left\lvert X_2\right\rangle, \left\lvert X_3\right\rangle$	$\left\lvert X_0\right\rangle = -2.98407\left\lvert X_0\right\rangle + 0.134274\left\lvert X_2\right\rangle - 0.0000324573\left\lvert X_3\right\rangle$
III	$\left\lvert X_0\right\rangle, \left\lvert X_1\right\rangle, \left\lvert X_2\right\rangle, \left\lvert X_3\right\rangle$	$\left\lvert Y\right\rangle^{III} = -2.94395\left\lvert X_0\right\rangle + 0.0633549\left\lvert X_1\right\rangle - 0.112056\left\lvert X_2\right\rangle$ $-0.0000363728\left\lvert X_3\right\rangle$

However, conceptually, SPECTRAL-SAR achieves a degree of novelty with respect to normal QSAR though that the spectral equation is given in terms of vectors rather than variables. Such features marks a fundamental achievements since this way we can deal at once with whole available data (of activity and descriptors) within a generalized vectorial space. Consequently, we may also use the spectral norm of the activity,

$$\left\| Y\right\rangle^{MEASURED/}_{PREDICTED} \right\| = \sqrt{\sum_{i=1}^{N} (y_i^2)^{MEASURED/}_{PREDICTED}} \, , \tag{1.30}$$

as the general tool by means which various models can be compared no matter of which dimensionality and of which multi-linear degree since they all reduce to a single number. This could help fulfill QSAR's old dream of providing a conceptual basis for the comparison of various models and end points by becoming a true science. Even more, while also accurately reproducing the statistics of the standard QSAR, the actual SPECTRAL-SAR permits the introduction of an alternative way of computing correlation factors by using the above spectral norm concept. As such the so called algebraic SPECTRAL-SAR correlation factor is defined as the ratio of the spectral norm of the predicted activity versus that of the measured one:

$$r_{S-SAR}^{ALGEBRAIC} = \frac{\left\| Y\right\rangle^{PREDICTED} \right\|}{\left\| Y\right\rangle^{MEASURED} \right\|} \, , \tag{1.31}$$

Applying equation (1.31) to the present case of the measured spectral norm of *Tetrahymena pyriformis* activity $\left\| Y\right\rangle^{MEASURED} \right\| = 6.83243$ the algebraic SPECTRAL-SAR correlation factors for the actual predicted models are given in Table 1.6 along the individual spectral norm of activity and the standard statistical correlation factor values.

Table 1.6. The predicted spectral norm, the statistic, and the algebraic correlation factors of the SPECTRAL-SAR models of Table 1.5, computed upon the general equations (1.30), (1.21), and (1.31) since the entry data of Table 1.3 are employed, respectively.

	Ia	Ib	Ic	IIa	IIb	IIc	III
$\left\| Y \right\rangle^{PREDICTED} \right\|$	3.86176	6.22803	6.0607	6.24858	6.32297	6.43641	6.44557
$r_{S-SAR}^{STATISTIC}$	0.53905	0.90759	0.88193	0.91074	0.92214	0.9395	0.9409
$r_{S-SAR}^{ALGEBRAIC}$	0.56521	0.91154	0.88705	0.91455	0.92543	0.94204	0.94338

The findings in Table 1.6 are twice relevant: first, because it is clear that the spectral norm parallels the statistic correlation factor; second, because, since the introduced algebraic correlation factor does the same job, it poses slightly higher values on a systematic basis.

In other words, one can say that in an algebraic sense the SPECTRAL-SAR furnishes systematically higher correlation factors than the standard QSAR does. This feature is also depicted in Figure 1.4 from where it is also noted that both correlation factors tend to approach each other near the ideal correlation factor, that is in the proximity of $r = 1.00$.

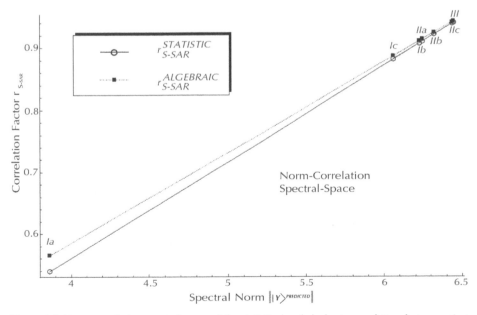

Figure 1.4. Norm correlation spectral space of the statistical and algebraic correlation factors against the spectral norm of the predicted SPECTRAL-SAR models of Table 1.6, respectively.

Nevertheless, we should note at this point that while a certain model does not satisfy the correlation factor criteria for being validated, i.e., $r > 0.84$, as is the case of the model (*Ia*) when only hydrophobicity is taken into account, this does not mean that the descriptor or chemical domain is less relevant; it is merely an indication that this descriptor may be further considered in a multivariate combination with others until produce better model.

Indeed, both within standard QSAR and SPECTRAL-SAR approaches all models except (*Ia*) are characterized by relevant statistics.

Next, aiming to see whether the obtained models can provide us a mechanistic model of chemical–biological interaction of tested xenobiotics on *Tetrahymena pyriformis* species, the introduced spectral norm is employed in conjunction with algebraic or statistic correlation factors to compute the *spectral paths* between these models. Such an endeavor may lead to an intra-species analysis of models and form the first step for designing of integrated test batteries (or an expert system) at the inter-species level of ecotoxicology.

In this respect, Table 1.7 presents the computed spectral distance between the models of the measured $Log(1/IGC_{50})$ endpoint of Table 1.3 though considering all path combinations that contain a single model for each class, with one and two descriptors, toward the closest model, that is (*III*), with respect to the ideal one. It follows that the paths are grouped according to the intermediary passing model while extreme models (initial and final) are kept fixed. Such ordered paths can be rationalized since a selection criterion is further introduced. Since paths are involved, one may learn from the well-established principle of nature according to which the events are linked by closest paths (in all classical and quantum spaces).

Table 1.7. Synopsis of the statistic and algebraic values of paths connecting the SPECTRAL-SAR models of Table 1.5 in the norm-correlation spectral-space of Figure 1.4.

Path	Value	
	Statistic	**Algebraic**
Ia-IIa-III	2.61485	2.61132
Ia-IIb-III	2.61485	2.61132
Ia-IIc-III	2.61485	2.61132
Ib-IIa-III	0.220072	0.219855
Ib-IIb-III	0.220072	0.219855
Ib-IIc-III	0.220072	0.219855
Ic-IIa-III	0.389359	0.388969
Ic-IIb-III	0.389359	0.388969
Ic-IIc-III	0.389359	0.388969

Therefore, we may formulate the *SPECTRAL-SAR least path principle* as follows: the hierarchy of models is driven by the minimum distance between endpoints (predicted norm of activities) of different classes of descriptors and of their combinations;

whenever multiple minimum paths are possible, that principle applies iteratively downwards between individual intermediate models of paths, starting with that one with minimal spectral norm.

In our case, according to the enounced minimum spectral path rule, the diagram of Figure 1.5 is constructed. It emphasizes different mechanistic hierarchies of the *Tetrahymena pyriformis* toxicophores. It comes out that, for instance, while three minimum paths result from Table 1.7, namely *Ib-IIa-III*, *Ib-IIb-III*, and *Ib-IIc-III*, only one is selected as giving the primary hierarchy, *Ib-IIa-III*, based on the fact that the spectral norm of *IIa* is the closest one to *Ib*. This is a purely mechanistic result since the correlation order in Table 1.6 would require that *IIc* be the next model chosen when starting from model *Ib*. At this point, we see that what is ordered from a statistical point of view may be degenerate in path length between the spectral norms. Therefore it appears that statistics might not be the most adequate criterion for SAR validity, since models with different correlations factors may be equally inter-related through spectral norms. Used exclusively, the statistic criteria will give little information about the subsidiary inter-species correlations in a unitary picture. On the contrary, the spectral path rule is able to formulate a scheme of connected paths between the models employing the natural principle of minimal action. *Minimal action here means that minimal length between spectral norms of different categories of endpoints is more favorable and comes firstly into a process driven by the succession of activities.* Thus, once the path *Ib-IIa-III* is naturally selected as the primary hierarchy of the ecotoxicity mechanism of *Tetrahymena pyriformis*, one can expect that, in this interpretation of the minimum spectral paths, the envisioned sequence of actions toward the measured one can be causally modeled as the action of polarizability followed by that of hydrophobicity and finally by that of total energy, through the optimization of molecular geometry during the chemical–biological interactions involved. This picture tells that the covalent interaction is the most dominant one, in this case, and drives the approach between the xenobiotics and the cells of organism; then enters into action the transfer through cellular membrane and finally the stabilization being assured by the stereo-specificity of the compounds linked to the receptor site. This way, a molecular mechanism may be coherently formulated in terms of norms of actions and of their inter-distances.

Whenever the primary route is inhibited, the second hierarchy of action follows by excluding the models previously involved and based on the same least principle of action. The second initial model will be chose that which is nearest to the first one on the spectral norm scale. Then, from all equivalent paths the next step is made toward the closes neighbor in the spectral norm sense.

The second hierarchy results along the endpoints path *Ic-IIb-III*, see Figure 1.5. This tells us that, by some subsidiary, slower action, the stereo-specificity selection is the first stage of the chemical–biological interaction analyzed, followed by membrane transport and only then by the stabilization of chemical bonds through polarizability.

If the secondary route is somehow repressed, as well the third way of ecotoxicological action of *Tetrahymena pyriformis* is also revealed as in Figure 1.5, *Ia-IIc-III*, again on the minimal activity action grounds constructed.

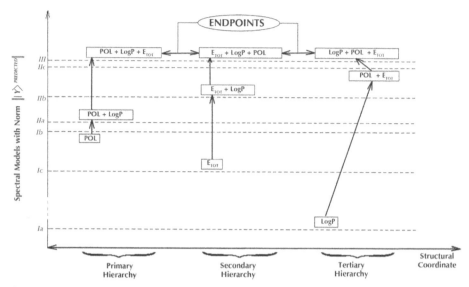

Figure 1.5. Spectral-structural models, designed through the rules of minimal SPECTRAL-SAR paths of Table 1.7, emphasizing the primary, secondary, and tertiary hierarchies forward the endpoints of the *Tetrahymena pyriformis* ecotoxicological activity according with data of Table 1.3, SPECTRAL-SAR equations of Table 1.5, and of the associated spectral norms computed upon equation (1.30).

It is not surprising that the application of minimal action principles on the spectral activity norms furnished many, however ordered, ways in which chemical–biological interaction are present in nature. This is in accordance with the heuristically truth that the nature reserves the privilege to develop many paths to achieve an action. The present SPECTRAL-SAR approach gives these new possibilities of hierarchically modeling of activities, in a way that the statistical analysis appears to be limited to single choices. Nevertheless, further work has to be performed by employing SPECTRAL-SAR method and of its minimal spectral path principle on many species and class of compounds in order to better validate the present results and algorithm.

CONCLUSION

Aiming to solve part of the many challenges posed by QSAR and its applications, with a view to generating a mechanistic-causal vision of the data recorded (measured or computed), this chapter introduces both a new analytical SAR modeling algorithm (the so called SPECTRAL-SAR method, idiomatic from Special Computing Trace of Algebraic SAR) and its associated minimum spectral action principle, following the activity norm of the models generated. As such, four possible branches of a QSAR expertise were identified, namely those based on the so called classical (of Hansch type), 3-dimensional (of CoMFA or MTD type), decisional (of genetic algorithm type), and orthogonal (of PCA type)—all proposing to furnish an appropriate analytical model for structure-chemical property or biological activity correlations. In this context the orthogonality problem was especially addressed, though the considered descriptors have to be as little collinear as possible in order to eliminate redundancies. Despite the fact that many QSAR approaches make use of algorithms that separate or transform initial non-orthogonal data into an orthogonal space, in search of a better correlation, many of them provide no significant improvement over the standard QSAR least square recipe. Instead, the present endeavor puts forth the orthogonal space (in Gram-Schmidt sense) only as an intermediate one in order to obtain from it, the spectral expansion of concerned activity and descriptors like vectors in a high-dimensional space.

This way, through more algebraic transparent transformations the spectral structure-activity relationships (SPECTRAL-SAR) are formulated as viable alternative to the previous standard QSAR method. The actual SPECTRAL-SAR approach also provides the framework in which the spectral norm can be formulated as assigning a single number to any SAR problem with the meaning of encoded of all information of a model, including the statistics. However, the spectral norm permits the spectral formulation of the minimal action principle applicable among various tested models. As such, the ecotoxicology of the *Tetrahymena pyriformis* was studied in detail providing the hierarchical paths of molecular actions toward the recorded activity. Since all consecrated criteria of a valid SAR analysis to an ecotoxicology study were included, the present added principle, in terms of minimum path over spectral norms of possible models for a certain set of data, unfolds the perspective of a real mechanistic interpretation of the chemical–biological interaction based on QSAR equation.

Nevertheless, further inter-species studies as well as the time-version of the least spectral norm principle will be in next undertaken in order to better reveal the features and advantages of the present SPECTRAL-SAR method.

KEYWORDS

- **Ecotoxicology**
- **Gram-Schmidt algorithm**
- **QSAR approaches**
- **QSAR equation**
- **Spectral paths**

Chapter 2

SPECTRAL-SAR Approach of the Enzymic Activity

Mihai V. Putz and Ana-Maria Putz

INTRODUCTION

In the context of the new spectral approach of the structure-activity relationship [see *Chapter 1 of this volume*], the enzymatic activity in terms of Hansch expansion is explored for predicting the equation of the hydrolysis velocity of the acetic acid esters with acetyl-cholinesterase. It follows that the actual spectral algorithm gives a similar prediction respecting the traditional quantitative structure-activity regression analysis so offering the key to replace it, when necessarily. Enzymes and enzymatic reactions are from the first researches in biochemistry of great interests due to their major role in catalysis, metabolism, but also from the point of view of inhibitors, drugs, and proteins [1].

From the catalytically standpoint the most general scheme of the enzymatic reactions can be described by the famous Michaelis–Menten mechanism [2],

$$E + S \underset{k_{-1}}{\overset{k_1}{\rightleftharpoons}} ES \xrightarrow{k_2} E + P \tag{2.1}$$

the reversible reaction between an enzyme E and a substrate S, involving an intermediary enzyme-substrate complex ES, with the rates k_1 and k_{-1}, which irreversibly gives the product P with the rate k_2.

However, the intensity of the enzymatic action is quantitatively determined by computing the variation in time of the concentration of any of the members of the equation (2.1), the result reflecting a sort of "molecular activity," that is the number of molecules or bonds modified in unit time.

Going to have an analytical expression for the enzymatic activity, the heuristic procedure employs the velocity of product formation Michaelis–Menten equation [3],

$$\frac{d}{dt}[P] = k_2 [E_{initial}] \frac{[S]}{[S] + K_M} \cong \frac{k_2}{K_M} [E_{initial}][S] \tag{2.2}$$

within a regime in which a substrate concentration [S] is sensible lower than that of the Michaelis constant $K_M = (k_{-1} + k_2)/k_1$, see the Appendix for simple derivation.

When, from equation (2.2), only the enzyme's catalytic efficiency k_2 / K_M is retained the so called biological activity may be introduced as [4]:

$$A = \log\left(\frac{k_2}{K_M}\right) \tag{2.3}$$

Still, at the same time, enzymes posse high specificity, a characteristic discovered in 1894 by Emil Fisher, being intensively studied for the role that they induce in medicinal chemistry, drug design, and ecotoxicology [5].

Therefore, the enzyme specificity is doubled by the mode of action and by the type of substrate, respectively.

In such, the enzymatic activity (2.3) cannot be expressed from (2.2) in a simple way.

This because the Michaelis–Menten approach of enzymatic catalysis is just an *in vitro* approximation of the enzymatic processes [6], being noted that in the course of transposing of the enzymes from *in vivo* to *in vitro* conditions some steric characteristic are lost [7].

More, there are some few but important enzymes that are not subscribed to the above Michaelis–Menten kinetic, even *in vitro*. An eminent example stands the blood cholinesterase of which activity grows to some maximum, depending on the substrate concentration, to decrease afterwards displaying so far a sigmoidal kinetic diagram [8].

Fortunately, there exist an alternative to Michaelis–Menten equation (2.2) to express the enzymatic activity (2.3), namely by performing quantitative correlation with the structural parameters of the binding molecules with which the concerned enzyme reacts to generate the studied biologically activity.

Basically, such procedure is circumvented to the consecrated quantitative structure-activity relationships (QSARs) that prescribe the activity expansion under the generic form:

$$A = B_0 + B_1 \left(\frac{electronic}{parameter} \right) + B_2 \left(\frac{hydrophobic}{parameter} \right) + B_3 \left(\frac{steric}{parameter} \right) + ... \qquad (2.4)$$

being this series known with the coefficients since the data sets of biological activities and those of the involved molecular parameters are analytically correlated.

However, when the series (2.4) is cut off after the first four coefficients the classical Hansch QSAR form equation is obtained [9].

Nevertheless, beside the already traditional regression methods in determining the correlation coefficients of (2.4) [10], a most recent spectral orthogonal method is employed in what next [11], and discussed from the statistical point of view, concerning the theoretical prediction of the enzymatic biological activity.

SYNOPSIS OF SPECTRAL-SAR METHOD

Without enter into details, see ref. [11] and Chapter 1 of this volume, if one has to solve correlation between a set of biological activities of N-compounds with the set of M-structural properties of each of them, a spectral algebraic algorithm can apply as follows.

First, the N-biological activities as well as their respective predictor variables are grouped in vectors, as displayed in Table 2.1, to which the additional unity vector $|X_0\rangle = |1 \ \ 1 \ \ ... \ \ 1\rangle$ was added.

Then, the searched correlation equation takes the spectral decomposition of the activity vector $|Y\rangle$ on the vectorial base $\{|X_0\rangle, |X_1\rangle, ..., |X_k\rangle, ..., |X_M\rangle\}$ with the form:

$$|Y\rangle = B_0|X_0\rangle + B_1|X_1\rangle + ... + B_k|X_k\rangle + ... + B_M|X_M\rangle. \tag{2.5}$$

Table 2.1. Synopsis of the SPECTRAL-SAR study descriptors.

Activity	Structural Predictor Variables					
$|Y\rangle$	$|X_0\rangle$	$|X_1\rangle$	∞	$|X_k\rangle$	∞	$|X_M\rangle$
y_1	1	x_{11}	∞	x_{1k}	∞	x_{1M}
y_2	1	x_{21}	∞	x_{2k}	∞	x_{2M}
∂	∂	∂	∂	∂	∂	∂
y_N	1	x_{N1}	∞	x_{Nk}	∞	x_{NM}

The general equation (2.5) is explicitly unfolded since the determinant equation

$$\begin{vmatrix} |Y\rangle & B_0^* & B_1^* & \cdots & B_k^* & \cdots & B_M^* \\ |X_0\rangle & 1 & 0 & \cdots & 0 & \cdots & 0 \\ |X_1\rangle & r_0^1 & 1 & \cdots & 0 & \cdots & 0 \\ \vdots & \vdots & \vdots & \vdots & & \vdots & \\ |X_k\rangle & r_0^k & r_1^k & \cdots & 1 & \cdots & 0 \\ \vdots & \vdots & \vdots & \vdots & & \vdots & \\ |X_M\rangle & r_0^M & r_1^M & \cdots & r_k^M & \cdots & 1 \end{vmatrix} = 0 \tag{2.6}$$

is solved respecting $|Y\rangle$ [11], where the involved parameters are given with the scalar product formulas:

$$r_i^k = \frac{\langle X_k | X_i^* \rangle}{\langle X_i^* | X_i^* \rangle}, \quad B_k^* = \frac{\langle X_k^* | Y \rangle}{\langle X_k^* | X_k^* \rangle}, \quad k = \overline{0, M}, \tag{2.7}$$

while, the orthogonal basis $\{|X_0^*\rangle, |X_1^*\rangle, ..., |X_k^*\rangle, ..., |X_M^*\rangle\}$ was introduced according with the rules:

$$|X_0^*\rangle = |X_0\rangle, \quad |X_k^*\rangle = |X_k\rangle - \sum_{i=0}^{k-1} r_i^k |X_i^*\rangle, \quad k = \overline{0, M}. \tag{2.8}$$

From this algorithm, there is clear that no statistical tool was invoked hitherto thus furnishing in principle an exact unequivocal structure-activity equation with a high predictable character. Such a methodology will be subsequently applied to study the enzymatic activity in a particular case of acetylcholinesterase.

RESULTS ON ACETYLCHOLINESTERAS (ACHE) ACTIVITY

As a working example, those enzymes which preferentially hydrolyze acetyl esters such as acetylcholine (ACh), called acetylcholinesterase (AChE) or acetylcholine acetylhydrolase (EC 3.1.1.7), are well studied due their important physiologic functions [7].

For instance, the acetylcholine, a combination of the ester of acetic acid with choline, $CH_3COOCH_2CH_2N + (CH_3)_3$, was the first identified neurotransmitter by Henry Hallett Dale and Otto Loewi in 1914.

They had proved its important role as a chemical transmitter in both the central and parasympathetic nervous system in humans and many other organisms and won the Nobel Prize in Physiology or Medicine in 1936, for these studies.

Acetylcholine can have some remarkable effects when administered directly, or when blocked, hindered, or mimicked; cataract surgery sometimes requires the use of acetylcholine to squeeze the pupil rapidly [12].

More, in most circumstances, acetylcholine is detached quickly after acting by the enzyme acetylcholinesterase.

Cholinesterase inhibitors delay the degradation of acetylcholine, and by acting on the central nervous system are used to reverse muscle relaxants, to treat Alzheimer's disease, having also a place in cardiopulmonary resuscitation [13].

Therefore, having a theoretical prediction for the biological activity of the cholinesterase when hydrolyzes with acetic acid esters (CH_3COOR) in terms of structural parameters of the R—radicals stands indeed as a most useful tool in chemical medicine.

To accomplish such functional relation the present SPECTRAL—SAR approach is employed to the experimental data of the studied AChE catalyzed hydrolysis rates (A_i) of the acetic acid esters' R-radicals, characterized by structural parameters as the Taft electronic withdrawing power constants (σ_i^*), the Hansch hydrophobicity parameters (π_i) and by the steric constants ($E_{s,i}$) that characterize the inhibition by steric crowding of the reaction center, respectively, collected in Table 2.2 accommodated in accordance with the Table 2.1 vectorial prescriptions.

Table 2.2. SPECTRAL-SAR synopsis of the enzymatic activities A_i of the hydrolysis of the cholinesterase enzyme with acetic acid esters jointly with the structural parameters of the esters' R- radicals: σ_i^*—Taft constants, ϖ_i—Hansch hydrophobicities, and $E_{s,i}$—steric constants, respectively [14].

Compound Formula (CH_3COOR), R—radical	A_i	$\lvert 1 \rangle$	σ_i^*	π_i	$E_{s,i}$
	$\lvert Y \rangle$	$\lvert X_0 \rangle$	$\lvert X_1 \rangle$	$\lvert X_2 \rangle$	$\lvert X_3 \rangle$
$-C_6H_5$	6.72	1	0.6	2.13	−0.38
$-CH_2CH_2C(CH_3)_3$	6.3	1	0	2.98	−0.4
$-CH_2CH_2SCH_2CH_3$	5.4	1	0.22	1.95	−0.44
$-CH_2SCH_2CH_3$	5.35	1	0.56	1.45	−0.44
$-CH_2CH_2CH(CH_3)_2$	5.32	1	0.	2.3	−0.43

Table 2.2. *(Continued)*

Compound Formula (CH$_3$COOR), R—radical	A_i / $\lvert Y \rangle$	$\lvert 1 \rangle$ / $\lvert X_0 \rangle$	σ_i^* / $\lvert X_1 \rangle$	π_i / $\lvert X_2 \rangle$	$E_{s,i}$ / $\lvert X_3 \rangle$
–CH$_2$CH$_2$NO$_2$	5.2	1	0.62	1.31	–0.4
–CH$_2$CH$_2$Cl	5.02	1	0.39	1.39	–0.48
–CH$_2$C≡CH	4.81	1	0.55	0.94	–0.6
n – C$_5$H$_{11}$	4.74	1	0.	2.5	–0.4
n – C$_7$H$_{13}$	4.75	1	0.	3.5	–0.4
-(CH$_2$)$_4$SCH$_2$CH$_3$	4.73	1	0.03	2.95	–0.4
cyclo – C$_6$H$_{11}$	4.71	1	0.	2.51	–0.98
n – (CH$_2$)$_4$H$_9$	4.72	1	0.	2.	–0.4
n – C$_6$H$_{13}$	4.68	1	0.	3.	–0.4
–CH$_2$CH$_2$CH$_2$SCH$_2$CH$_3$	4.67	1	0.08	2.45	–0.4
–CH$_2$C$_6$H$_5$	4.66	1	0.25	2.26	–0.38
–CH$_2$CH(CH$_3$)$_2$	4.32	1	0.	1.8	–0.35
–CH$_2$CH=CH$_2$	4.1	1	0.23	1.23	–0.2
n – C$_3$H$_7$	3.91	1	0.	1.5	–0.39
–CH(CH$_3$)CH$_2$CH$_3$	3.69	1	0.	1.8	–0.93
–CH$_2$CH$_3$	3.36	1	0.	1.	–0.36
–CH$_3$	3.	1	0.2	0.5	–0.07
–CH(CH$_3$)$_2$	2.72	1	0.	1.3	–0.93
–C(CH$_3$)$_3$	1.3	1	0.	1.98	–1.74
–C(CH$_3$)$_2$CH$_2$CH$_3$	1.3	1	0.	2.48	–1.74

Thus, the actual study is performed within a classical Hansch cut off of the relation (2.4). In these conditions, the SPECTRAL–SAR functional relationship is abstracted from the equation:

$$
\begin{vmatrix}
\lvert Y \rangle & 4.3792 & 2.51236 & 0.939019 & 1.93529 \\
\lvert X_0 \rangle & 1 & 0 & 0 & 0 \\
\lvert X_1 \rangle & 0.1492 & 1 & 0 & 0 \\
\lvert X_2 \rangle & 1.9684 & -1.47315 & 1 & 0 \\
\lvert X_3 \rangle & -0.5616 & 0.479246 & -0.0150042 & 1
\end{vmatrix} = 0
\qquad (2.9)
$$

written upon the previous exposed methodology, see equations (2.6)–(2.8), to give out the type-expansion (2.5) under the analytical specialization:

$$
\lvert Y \rangle = 3.1113 \lvert X_0 \rangle + 3.01097 \lvert X_1 \rangle + 0.968056 \lvert X_2 \rangle + 1.93529 \lvert X_3 \rangle
\qquad (2.10)
$$

Finally, recognizing the vectorial correspondences with the studied parameters in Table 2.2 the actual spectral Hansch-SAR is obtained as:

$$\overset{\cap}{A_i} = 3.1113 + 3.01097\sigma_i^* + 0.968056\pi_i + 1.93529E_{s,i} \tag{2.11}$$

Obviously, in deriving (2.11) no statistical procedures were involved. Nevertheless, the robustness of the above AChE activity SPECTRAL-SAR will be discussed in what follows.

Worth checking whether the spectral SAR given by equation (2.11) stands as a good model for the biological activities of Table 2.2 so that to can be adopted with a predictable value.

To do that, a series of goodness indicators may be used among which we will explore the standard error of estimate (*SEE*), the correlation coefficient (*r*) and the Fisher test ($F_{f_1=M, f_2=N-M-1}$) [15].

First, the total sum of squares (*SQ*), measuring the dispersion of the working activities around their average ($\overline{A} = (1/N)\sum_{i=1}^{N} A_i$),

$$SQ = \sum_{i=1}^{N} \left(A_i - \overline{A} \right)^2, \tag{2.12}$$

together with the total sum of the residues (*SR*), measuring the spreading of the input activities respecting their estimated counterparts,

$$SR = \sum_{i=1}^{N} \left(A_i - \overset{\cap}{A_i} \right)^2, \tag{2.13}$$

are calculated with the help of equation (2.11) and of the data from the Table 2.2 to give $SQ = 39.6149$ and $SR = 8.76532$, respectively.

Then, with these quantities the concerned indicators, *SEE*, *r*, and $F_{f_1, f_2;\alpha}$, are computed to yield:

$$SEE = \sqrt{\frac{SR}{N-M-1}} = 0.646062, \tag{2.14}$$

$$r = \sqrt{1 - \frac{SR}{SQ}} = 0.882461, \tag{2.15}$$

$$F_{f_1=M, f_2=N-M-1} = \frac{N-M-1}{M}\left(\frac{SQ}{SR} - 1\right) = 24.6365. \tag{2.16}$$

Inspecting the values of the indicators (2.14), (2.15) is observed that even the *SEE* and *r* display rather modest output the cutting judgment is giving by the calculated Fisher index (2.16).

Applying the Fisher line of investigation, appears that the computed index (2.16) is more than four times higher than the corresponded tabulated one for over 99% trusting test ($F_{3,21;99\%} = 4.87$) [15] results that the actual SAR equation (2.11) represents with larger than 99% confinement the goodness of the modeled biological data respecting their simple average prediction.

Remarkably, the very same results were founded out previously by the QSAR group of Timișoara, engaging the multivariate regression [4, 7, 9].

Therefore, due to the fact that the present approach recover similar conclusions on completely different non-statistical way, results that the SPECTRAL—SAR analysis demonstrated its reliability in predicting biological activity, here indicating at the quantitative level that the strong electronic withdraw as well as the increasing hydrophobicity of R-radicals favor the reaction, while the more negative E_s values for R-radicals lower the enzymatic reaction rate.

However, despite its similarity in results with the traditional multivariate statistical methods, for example, the principal component analysis, cluster analysis, discriminant analysis, partial least squares regression, or artificial neural networks [10], the actual employed SPECTRAL SAR method has the strength of no dependency on the way in which the input data are considered thus being less dependent on the outliers detection and more direct.

Further applications of SPECTRAL-SAR method on enzyme-substrate specificity and ecotoxicology may be performed as well and will be reported in subsequent studies.

CONCLUSION

In the continuous efforts to implement quantitative structure-activity relationship models to predict biological specific interactions, mutagenicity, or toxicity of large series of compounds, the spectral non-statistical algorithm was tested in determination of the acetylcholinesterase catalyzed bimolecular hydrolysis rate equation within classical Hansch structural expansion of acetic acid esters' parameters.

It was found out that the actual procedure gives reliable results *per se* and being in a similarity relation with the alternative statistical methods.

There is therefore hope that some of the mathematical limits and problems of selections involved in the current regression analysis will be removed following the pure algebraic way employing the vectorial spaces and the coordinate transformation across orthogonal bases, toward a generalized quantum biochemical space.

This is, at the end, the true meaning of the quantitative correlation between the intrinsic chemical reactivity and the manifested biological activity.

APPENDIX: MICHAELIS–MENTEN ENZYME KINETICS

When the law of mass action is considered for the reaction (2.1), the time evolution scheme can be draw as the system of the coupled nonlinear differential equations [6, 16]:

$$\frac{d}{dt}[S] = -k_1[E][S] + k_{-1}[ES] \tag{2.A1a}$$

$$\frac{d}{dt}[E] = -k_1[E][S] + (k_{-1} + k_2)[ES] \tag{2.A1b}$$

$$\frac{d}{dt}[ES] = k_1[E][S] - (k_{-1} + k_2)[ES] \tag{2.A1c}$$

$$\frac{d}{dt}[P] = k_2[ES] \tag{2.A1d}$$

with initial conditions $([S],[E],[ES],[P]) = ([S_0],[E_0],0,0)$ at the time $t=0$.

The set of equations (A1) can be simplified in three steps.

First, it can be seen that when the equations (2.A1b) and (2.A1c) are added, the conservation law for enzyme is obtained:

$$[E](t) + [ES](t) = [E_0] \tag{2.A2a}$$

while the combination of equations (2.A1a), (2.A1c), and (2.A1d) leads to the conservation law for the substrate:

$$[S](t) + [ES](t) + [P](t) = [S_0] \tag{2.A2b}$$

With the help of identities (A2), the system of differential equations (A1) takes the reduced form:

$$\frac{d}{dt}[S] = -k_1[S]([E_0] - [ES]) + k_{-1}[ES] \tag{2.A3a}$$

$$\frac{d}{dt}[ES] = k_1[S]([E_0] - [ES]) - (k_{-1} + k_2)[ES] \tag{2.A3b}$$

in terms of substrate and substrate enzyme concentrations only, $[S]$ and $[ES]$, respectively.

Then, employing the *in vitro* conditions, the enzyme can always be saturated with the substrate, so that the *quasi-steady-state* (or equilibrium) *approximation* (QSSA) may apply to the intermediate formed complex in (2.1). It implies imposing on (2.A3b) the mathematical constrain [17–19]:

$$\frac{d}{dt}[ES] \cong 0 \tag{2.A4}$$

yielding with its equivalent form:

$$[ES] = \frac{[E_0][S]}{[S] + K_M} \tag{2.A5}$$

where the reaction parameter

$$K_M = \frac{k_{-1} + k_2}{k_1}$$
(2.A6)

is known as the *Michaelis–Menten constant* [3].

Now, plugging relation (2.A5) into the equation (2.A3a), we get the decoupled differential equation for the substrate consumption rate:

$$\frac{d}{dt}[S] = -\frac{V_{max}[S]}{[S] + K_M}$$
(2.A7)

where

$$V_{max} = k_2[E_0]$$
(2.A8)

has been set as the *maximum velocity of reaction*.

At this point, the system (2.A1) achieves its minimum dimension consisting in one equation for the substrate concentration. However, by combining the equations (2.A1d) and (2.A5), the velocity of the product formation also comes out,

$$v = \frac{d}{dt}[P] = \frac{V_{max}[S]}{[S] + K_M}$$
(2.A9)

as the famous Michaelis–Menten equation [2, 3].

KEYWORDS

- Acetylcholinesterase
- Ecotoxicology
- Hydrolysis velocity
- Metabolism

PERMISSIONS

This Chapter was previously published in Ann. West University of Timisoara, Series of Chemistry, 15(2) (2006) 167-176 as A Spectral Approach of the Molecular Structure-Biological Activity Relationship Part II. The Enzymatic Activity (Mihai V. Putz and Ana-Maria Lacrama).

Chapter 3

Designing Ecotoxico-logistical Batteries by SPECTRAL-SAR Maps

Ana-Maria Putz, Mihai V. Putz, and Vasile Ostafe

INTRODUCTION

For the ecotoxicological risk assessment of chemicals the information regarding their toxicity is required from different trophic levels, from primer producers to secondary consumers. As a consequence, having a battery of test systems would be representative for a wide range of chemicals. The methodology and molecular implications in choosing and applying a battery test are reviewed. In numerical analysis, the recent spectral method of approaching the multiple linear functional correlations for quantifying the structural-activities relationships (SARs) is considered to predict the overall toxicity of the tested species. The toxic activities are implemented through the usually parameter $Log(1/EC_{50})$ and employed within a novel LOGISTICAL-SPECTRAL-SAR algorithm of generalized chemical–biological interactions providing a mechanistic heuristically over-view on the inter-species correlations. The present spectral method adds plus value in an ecotoxicological test battery for molecular toxic appraisal of new substances that have not been tested yet. Toxicity could be defined as a function of the ability of a chemical to reach the active site and its capacity to react covalently at the active site, for these toxic effects other than resulting from receptor-binding (e.g., endocrine disruption) [1]. The first phenomenon could be described by the hydrophobicity of a compound (the logarithm of the octanol-water partition coefficient, $LogP$) and the second one by a measure of electrophilicity (the energy of the lowest unoccupied molecular orbital, E_{LUMO}) [2].

In the case of new chemicals, we have to test their toxicity in all the surroundings that they could reach a destination: soil, fresh, and salt water. Knowing that not all-relevant target sites are found in a single organism, it is widely recognized that no single bioassay can be used to assess the toxic effects of every different mode of action. Therefore, a battery of test methods is required [3, 4]. Comparing the concentration in an environmental compartment with a predicted no-effect concentration we are performing the risk assessment of chemicals. An evaluation of the first EU priority list has found that two or more tests with soil organisms, as needed to derive a predicted no effect concentration (PNEC), being available for only about 35% of these high priority, high production volume chemicals [5]. Therefore, in the absence of test data with soil organisms, the technical guidance document (TGD) suggests use of the equilibrium partitioning method to extrapolate soil toxicity from the aquatic toxicity [5, 6].

Risk assessment has to consider the mobility and bioavailability of the contaminants, and their uptake and metabolization by microorganisms, plants and animals including man. For an appropriate consideration of all these factors, biotests have to be applied which reflect the pollutant-induced reduction of the habitat function [7] and the retention function [8] of soil or soil material. To consider both functions, a battery of evaluated biotests is obligatory, which integrates all adverse effects caused by contaminated soil [9]. A minimum ecotoxicological test battery should at least include bacteria, vegetables, invertebrates, and mammalian and non-mammalian cells [3].

A good example of an ecotoxicological test battery was found by Repetto G., Jos A., and his colleagues in 2001. The battery included the tests: the immobilization of *Daphnia magna*, bioluminescence inhibition in the bacterium *Vibrio fischeri*, growth inhibition of the alga *Chlorella vulgaris*, micronuclei induction and root growth inhibition in the plant *Allium cepa*; cell morphology, total protein content, MTS metabolization, lactate dehydrogenase leakage activity and glucose-6-phosphate dehydrogenase activity were studied in the salmonid fish cell line RTG-2; the total protein content, LDH activity and MTT metabolization were investigated in Vero monkey kidney cells [3].

Assessment of potential impacts in the receiving water to which an effluent is discharged requires inclusion in the test battery of species representing different trophic levels: typically algae, which are primary producers, invertebrates, primary consumers, and fishes, secondary consumers [4, 10].

Recently, another very useful strategy has been proposed for ecotoxicological evaluation, test-kit-concept, in which a major role is dedicated to the test battery ecotoxicity/toxicity tool, which has to allow a systematic evaluation of (eco) toxicological properties of chemicals on biological systems of different complexity [11]. The studies with test battery have to be performed at different levels: molecular, cellular, organism, and ecosystem. For testing the adverse effects of new compounds on environment, it will be easy (and cheaper) to start with the study of the effects at molecular level of these chemicals on some enzyme activities, selected by definite criteria that will allow the reductions of experiments. This can be realized studying the effect of that compound on some well chooses enzymes. To make a good enzymes kit for ecotoxicological test battery, the selected enzymes has to fulfill some requirements: to be part from a major biochemical pathway, to be found in almost all organisms and to be extracted/purified from a cheap biological source [1]. Multienzimatic test battery is a model for testing at molecular level the toxicity of chemical compounds, which was also applied by our group [12].

GENERALIZED CHEMICAL–BIOLOGICAL-KINETICS

Taking in consideration that the toxicity of the new chemicals depends of the calculated inhibitions constants, the way in which the new kinetics parameters associated with respectively enzymatic reaction are calculated, it is critical for a good practically prediction.

The mechanism of the biological activity produced by a substance usually involve the combination between the molecules of those substance, called effector or ligand (L) with a receptor (R), a protein, a biologically macromolecule, a complex of macromolecules from within of cell. The intensity of the biological action is illustrated in an ordinary way as logarithm of inverse of the concentration C which produces a specify biological answer, that is mean:

$$A = \text{Log}\ (1/C) \tag{3.1}$$

In many situations C_{50} concentration is used, that mean those molar concentration that produce 50% from the maximum biological activity. It could be shown that the biological activity, A, is proportional with the affinity of the molecule or of the ligand, L_i for the receptor, R, which stay at the basis of explicated biological action. The most rational hypothesis concerning the mode of action of the bioactive substances presumes that the biological activity produce by a ligand L is proportional with the complexation degree of the receptor R from the L. In this situation, the presuming a biological activity of a% (comparing from the maximum of 100%) is produced by a concentration, $[L]$ of ligand.

On the other side, it has long been recognized that the chemical reactions that support life are mediated by enzymes and their kinetics. Usually, enzyme catalysis is based on the reversible reaction between and enzyme E and a substrate S with rate constant k_1 to form an intermediate enzyme-substrate complex ES. The complex then reacts irreversibly with rate constant k_2, regenerating the enzyme E and producing product P.

Therefore, seeing together we may compare the ligand (L)—receptor (R) with substrate (S)—enzyme (E) through the items of Table 3.1 toward a generalization of the bio-kinetics.

Table 3.1. The face-to-face ligand-receptor (L-R) and substrate-enzyme (S-E) kinetics.

Property	L-R kinetics	S-E kinetics
Reaction	$R + L \overset{K}{\leftrightarrow} C^*$	$E + S_i \underset{k_{-1}}{\overset{k_1}{\leftrightarrow}} ES \overset{k_2}{\to} E + P$
Species	R	E
Chemicals	L	S
Equation	$\dfrac{\alpha}{100} = \dfrac{[L]}{[L]+K}$	$\dfrac{v}{V_{max}} = \dfrac{[S]}{[S]+K_M}$
Constant	$K = \dfrac{[R][L]}{[C^*]}$	$K_M = \dfrac{k_{-1}+k_2}{k_1}$
Graph		

Therefore, we may safety generalize the chemical-biological-kinetics in terms of the general rate of biological uptake β respecting the chemical concentration $[\chi]$

$$\beta = \frac{\beta_{max}[\chi]}{[\chi] + EC_{50}} \tag{3.2}$$

Recognizing in equation (3.2) the temporal link between the biological activity and chemical concentration stands as:

$$\beta = -\frac{d}{dt}[\chi](t) \tag{3.3}$$

the full temporal version of it can be widely formulated as:

$$-\frac{d}{dt}[\chi](t) = \frac{\beta_{max}[\chi](t)}{[\chi](t) + EC_{50}}. \tag{3.4}$$

However, the main problem with the equation (3.4) is that, it accounts only for the velocity of the initial time of the reaction. The information that is outside the first moments of the inherent progress curve is virtually lost or neglected.

Another complication of the equation (3.4) is that, even describing a generalized kinetic, differs than the ordinary chemical ones by its rectangular hyperbola shape, rather by an expected exponential form.

Instead, a further generalized kinetics may be assumed, which was recently applied to enzymic Michaelis–Menten case too, see [13] and the Appendix, namely:

$$-\frac{d}{dt}[\chi](t) = \beta_{max}\left(1 - e^{-\frac{[\chi](t)}{EC_{50}}}\right) \tag{3.5}$$

that is clearly reduced to the above equation (3.4) in the first order expansion of the chemical concentration time evolution respecting the 50-effect concentration (EC_{50}) observed.

Very impressive, although the original chemical-biological-kinetics (3.4) leads with no analytical solution the actual working kinetics (3.5) provides the so called logistical solution under the form, see [14] and the Appendix:

$$[\chi](t) = EC_{50}\ln\left(1 + e^{-\frac{\beta_{max}t}{EC_{50}}}\left(e^{\frac{[\chi_0]}{EC_{50}}} - 1\right)\right) \tag{3.6}$$

The reliability of the logistic solution (3.6) was previously tested on enzyme kinetics, within various mechanism, with remarkable results [14, 15] thus constituting a trusted background for employing it to the envisaged currently ecotoxicological studies.

ON THE USE OF THE CHEMICAL COMPOUNDS IN ECOTOXICOLOGY

Premises of QSAR in Ecotoxicology

As the chemical testing became more uneconomical and the quantitative structure-activity relationships (QSARs) models for predicting toxicity become more robust, there will be an emergency to develop and apply QSARs to predict both the human health and environmental effects of those chemicals [16].

There has been considerable interest in the prediction of the *in vitro* toxicity data using QSARs because, behind the advantage of easily data obtaining, often have direct mechanistic significance, in contrast to *in vivo* data that are usually result from the diverse mechanism of toxic action [17]. Consonant was recognized the necessity of qualitative toxicity data as the inputs for QSARs [10].

In ecotoxicology, the term human-surrogate species is used to describe those species (most often mammalian) that supply information to allow for insight to be made regarding human exposure to chemicals [16].

The ecologically relevant species indicate those species that may provide information that allow judgments to be made regarding the environmental effects of chemical [17]. In the case of ecotoxicity QSARs (e.g., the one developed for aquatic endpoints), the log-based continuous toxicological-data (e.g., LC_{50} or EC_{50}) and molecular descriptor data (e.g., quantum chemical and physicochemical parameters), are linked together by a statistical method such as regression analysis [18, 19].

Concerning the interactions between a ligand and a receptor, a number of QSARs and SARs methods are available to model this process. To predict which compounds may act in the same receptor, molecular similarity-based methods are applied to QSAR analysis. From chemicals representative of the diversity of structures for which prediction is to be made it is necessarily well-defined biological and toxicological data [20].

For a QSAR to have a mechanistically meaningful it must be based on physico-chemical descriptors that are right to the consideration of toxicity [1]. The inclination is inclusion of a large number of descriptors, which might improve the statistical fit of the model, but leads to false, unstable or invalid models, so the reduction of variable used in QSAR is needed [21, 22]. The more mechanistically transparent the QSAR, the more trusting the applicability domain can be defined. The comparison between the methods is complicated by the use of different descriptors and/or number of compounds in the same model.

Another problem was the lack of an agreed set of criteria for high quality QSARs. The choice of methods that we want to compare is restricted because not all available methods may be appropriate for a particular purpose, and application of standard approaches to different data sets could go against the principle of Occam's Razor, that is, to maintain and use the simplest model [10]. The development of QSARs on the basis of mechanism of action is often a difficult task. Thus, development of models that accurately predict toxicity without requiring the initial identification of an exact mechanism can be useful [22, 23].

Basic QSAR Descriptors in Ecotoxicology

Essentially, there are three main types of descriptors for use in the development of standard QSARs studies: hydrophobic, electronic, and steric descriptors [24, 25].

Steric descriptors are associated with the size and shape of molecules. Molecular size and their contours are important for many biological processes, like the ability of drug to bind at a receptor site. It is an important factor in the reactivity and metabolism of molecules as molecule aspects associated with reactivity and metabolism may be sterically hindered. Also it is an important factor in the ability of a chemical to pass through a membrane, because long thin molecules may be more able to permeate through the gaps between cells than more spherical molecules. Chirality could be also described and quantified using molecular shape [26]. The size and the bulk of a molecule can be quantified using descriptors like molecular weight and molecular volume [27]. Molecular shape which refers to the distribution of molecular bulk according to the conformation has been much less easy to quantify [28, 29].

There are two approaches to mutagenicity prediction, based on a mechanistic understanding of the process involved, qualitative identification of molecular substructures that account for electrophilic reactivity and quantitative description of the molecule in terms of its hydrophobicity or molecular orbital configuration or both [30].

$LogK_{ow}$ is the most used descriptor used for accounting of hydrophobicity but there is a broad variability in the type of descriptors used to account for electrophilicity and hydrogen-bonding ability [22]. Hydrophobicity and electrophilicity control the acute aquatic of non-specifically acting compounds [22, 31] and the magnitude of toxicity is determined by hydrophobicity and hydrogen-bonding capacity [22, 32].

QSARs for membrane permeability (QSPRs) are applicable for toxicological purposes and they will be applied in the prediction of risk assessment of chemicals. The ability of a compound to pass across a membrane is related with its ability to partition into and out of a relatively hydrophobic region. In modeling these events is assumed to passive diffusion. Permeability will be related to hydrophobicity, and $LogP$ will be the descriptor. Because there is also an upper limit on the size of the molecule that will pass through a membrane, molecular weight, or molecular bulk is needed [16]. Thus, the molecular size (as molecular weight) and hydrophobicity ($LogK_{ow}$) are the main determinants of transdermal penetration [33].

The toxic potency of aromatic molecules for aquatic toxicity endpoints can be modeled by two factors—hydrophobicity and stereo-electronic effects (i.e., reactivity, steric hindrance) [19]. The key descriptors for controlling the penetration ability of a compound are hydrophobicity, molecular size, and hydrogen-bonding ability [34].

The dimirystoyl phosphatidyl choline (DMPC)-water system has been proposed as an alternative to the 1-octanol-water system to assess the hydrophobicity of chemical substances; 1-octanol was recognized as a good surrogate organic phase for the modeling of membrane-water partitioning of chemicals. Other QSARs for the prediction of $LogK_{DMPC-W}$ not including $LogK_{ow}$, confirmed the role of molecular branching, size, and carbon content on the partition coefficient. Even if there are big differences between the molecular structures, $LogK_{DMPC-W}$ and $LogK_{ow}$, were found to be very firmly related, suggesting there is little advantage in the use of DMPC over 1-octanol as the non-polar phase in partitioning studies [35].

Guidelines for Development of Ecotoxicological QSARs

Schultz and Cronin have been identified the characteristics for ecotoxicological QSAR [36]: reliable ecotoxicological data, high quality, interpretable and reproducible descriptors of a number, and type consistent with the endpoint being modeled; and the used of a statistical procedure allowing the development of a rigorous and transparent mathematical model.

For establish validity and accuracy criteria for accepting (Q)SARs, guidance and guidelines were needed. Only if regulatory authorities and industry can agree on the acceptable number of false negative and false positive results, animal testing could be reduced. But the regulatory authorities are primarily concerned with false negative results because if a (Q)SARs predict too many false negatives, animal testing will be asked for. On the other hand, industry will ask for more animal testing if too many false positives will be. So the validity and the available tools in QSARs have to be explored further [37].

Following the same line, Bradbury and co-workers developed a review of conceptual approaches for derivation of QSAR for ecotoxicological effects of organic chemicals. They spoke about the tentative assumptions about the modes of action for narcotics that have evolved from general perturbation of cellular membranes due to a nonspecific partitioning of xenobiotics, to mechanisms that invoke partitioning into specific membrane microsites or hydrophobic pockets of membrane bound proteins [38].

Hermens and Verhaar review the four classes of chemicals.

Class I, represented by relatively unreactive chemicals with a nonspecific mode of action and they represent the base-line toxicity or nonpolar narcosis or simple narcosis [38, 39]. *Narcosis* is a reversible state of kept activity of protoplasmic structures, which is a result of exposure to the xenobiotic. Also the terms narcosis and general anaesthesia are used as equivalently in the circumstance of intact organisms. It presumed that, when the xenobiotic blood concentration is at equilibrium with the aqueous exposure concentration the aqueous concentrations of narcotic chemicals are proportional to the concentrations at the site of action [38]. In class I of chemicals we found aliphatic and aromatic (halogenated hydrocarbons) [39].

Class II of chemicals, *polar narcotics,* include: aromatics amine, nitroaromatics, anilines, phenols, and pyridines that do not have substituents associated with reactive or uncoupling modes of action. Polar narcotics act with a different molecular mechanism than baseline narcotics [38, 39].

Class III, the reactive chemicals, include epoxides, aldehydes, aziridines, quinones (generally, all alkylating agents) [39]. These chemicals (or their activated metabolites) react covalently with nucleophilic sites in cellular biomacromolecules (e.g., through nucleophilic substitution, Michael-type addition, or Schiff-base reactions) or gain an oxidative stress through redox cycling to derive toxic effects [38].

Class IV, the chemicals with a specific mode of action, includes organophosphates, chlorinated, and pesticides [39].

Also Bradbury and co-workers review the oxidative phosphorylation uncoupling QSARs, which embody these chemicals that are typically weak acids and are represent by phenols, anilines, and pyridines but, different from polar narcotics, include multiple electronegative groups (i.e., more than one nitro substituents, more than three halogen substituents, or both bonded to the aromatic ring), see Figure 3.1.

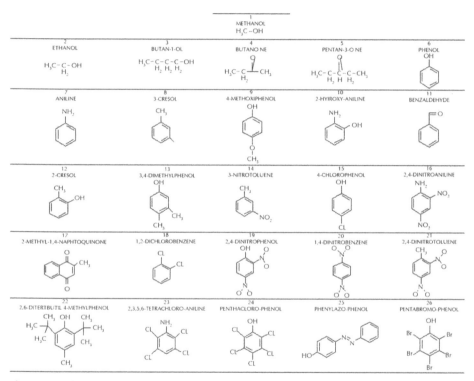

Figure 3.1. The series of chemical compound used for assessing toxicity at the level of biological species within the eco-battery modeling of the present study.

Bradbury said that a mode of action domain must be clearly defined in terms of the biological model, endpoint, and exposure contour [40]. The use of mode of action-based QSARs requires recognition of both: toxic mechanisms and the critical structural characteristics and properties of a chemical that rule its activity by a specific mechanism. The problem is that many chemical QSAR classes that historically were associated with a baseline narcosis include chemicals that act via a baseline narcosis mode of action, as well as chemicals that act through an electrophilic-based mode of action.

On the contrary, chemical classes not usually identified as acting by a *baseline narcosis* mode of action, such as the phenols, include chemicals that act through *baseline narcosis, polar narcosis, oxidative phosphorylation uncoupling,* or *electrophilic-based* modes of action [38].

One approach does develop QSAR equation is that of using congener series of chemicals. The initial research in the field of QSARs for ecological risk assessments predicated on the supposition chemicals from the same chemical class should behave in a toxicologically similar manner. Accordingly, homologous series of chemicals were used to develop structure-toxicity relationships and the premise was made that toxic effects were transmitting by common structural components used in chemical class task. Congener-derived QSARs are restricted because of the short structural domain of which they are grounded [36, 38].

Further, the attention was to development of QSARs on the basis of mechanisms of action, from narcosis to reactive mode of action that, nevertheless is often a difficult task [22].

The assumption was made that potency varied with chemical uptake, which correlated with the hydrophobicity of substituent moieties within the chemical class. Their opinion is that the power of chemicals that act through mechanisms other than narcosis is, in part, related to their partitioning into hydrophobic biological compartments, and also is a fundamental principle for QSARs. Thus, octanol-water partition coefficients are typically working in QSARs as the chemical descriptor that comprise that aspect of variability in toxicity across chemicals that is exclusively characteristic to varying degrees of hydrophobicity, and uptake, of xenobiotics [38].

ON THE USE OF THE BIOLOGICAL SPECIES IN ECOTOXICOLOGY

There is widely recognized that when assessing an ecotoxicological battery different species from various phylogeny has to be considered. Nevertheless, whishing to deliver a minimal yet valuable set of species to model a battery through the chemical–biological interaction with the chemical compounds of Figure 3.1, there will be here exposed the main eco-biological characteristics of *Chlorella vulgaris*, *Vibrio fischeri*, and *Pimephales promelas* accounting for an algae, a bacterium, and a fish species, respectively.

Chlorella vulgaris

Classified as Kingdom: *Protista*> Phylum: *Chlorophyta* > Class: *Chlorophyceae* > Order: *Chlorococcales* > Family: *Oocystaceae* > Genus: *Chlorella* [41, 42], the green algae, or *Chlorophyta*, compose the largest and the most varied phylum of algae.

They are characterized by their grassy green color, attributable to the pigments chlorophyll a and b, a, b, and c caratones, and several xanthophylls, and also by their storage of plant starch (a b 1,4-linked polyglucan). Among the algae the *Chlorophyta* are the most closely related to the plants based on their similar photosynthetic pigments, storage of plant starch, and the fine-structural organization of the chloroplast with the photosynthetic lamellae stacked together to give grana and intergranal regions [44].

Chlorella is a genus of single-celled green algae, belonging to the phylum Chlorophyta. It has a spherical shape, about 2–10 micrometer in diameter. It depends on photosynthesis for growth and multiplies rapidly, requiring only carbon dioxide, water, sunlight, and small amounts of minerals [45].

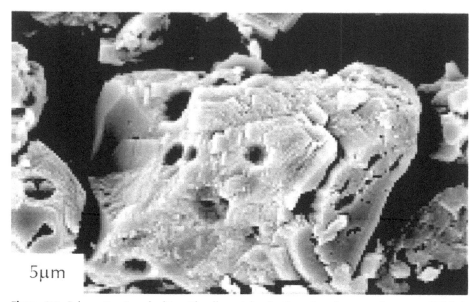

Figure 3.2. Eukaryotic pico-plankton Clorella as takes by Scanning-electron microscopy (SEM) *in vitro* [43].

Toxicity data, $Log(1/EC_{50})$ against the chemical compounds of Figure 3.1, for *Chlorella vulgaris*, brought from Cronin et al. (2004) [10], were determined in a biochemical assay utilizing the unicellular algae in their logarithmic phase of their growth cycle. The used temperature was between 25 and 30°C and the pH of the buffer solution was 6.9. Assays were done following the protocol described by Worgan et al. [46] with a 15 min static design. The disappearance of FDA (fluorescein diacetate) was accounted for by spectrofluorimetric measurement of fluorescein (the product of hydrolysis) [47] at an excitation wavelength of 465 nm and an emission wavelength of 515 nm [10].

Vibrio fischeri

Classified as Kingdom: *Monera* > Phylum: *Proteobacteria* > Class: *Gamma Proteobacteria* > Order: *Vibrionales* > Family: *Vibrionaceae* > Genus: *Vibrio*, the Species: *V. fischeri* is a rod-shaped bacterium found globally in the marine environments. It has bioluminescent properties and is found predominantly in symbiosis with various marine animals, such as bobtail squid, see Figure 3.3. It is heterotrophic and moves by means of flagella [42, 48(a)–(d)].

Free living *V. fischeri* survives on decaying organic matter. The bacterium is a key research organism for examination of microbial fluorescence and bacterial-animal symbiosis. Planktonic *V. fischeri* are found in very low quantities in almost all oceans of the world, found preferentially in temperate and sub-tropical waters. These free-living *V. fischeri* live on organics within the water. They are found in higher concentration in symbiosis with certain deep-sea marine life within special light-organs, or as a part of the normal enteral (gut) microbiota of marine animals.

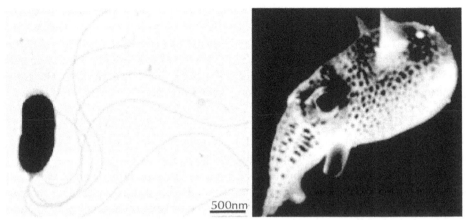

Figure 3.3. Left: *V. fischeri* on aqueous media as taken by *in vitro* electron micrographs [49] Right: *V. fischeri* on Hawaiian Bobtail Squid (Animalia > Mollusca > Cephalopoda > Sepiolida > Sepiolidae) [48(a)].

Additionally, the bacteria can be pathogenic to certain species of marine invertebrates, some of which are commercially farmed in aquaculture. This disease is known as luminous vibriosis. Symbiotic relationships in monocentrid fishes and sepolid squid appear to have evolved separately. The most prolific of these relationships is with the Hawaiian bobtail squid (*Euprymna scolopes*).

Free-living *V. fischeri* in the ocean waters inoculate the light organs of juvenile squid and fish. Ciliated cells within the light organs selectively tie in the symbiotic bacteria. These cells promote the growth of the symbionts and actively reject any competitors. Through quorum sensing the bacteria cause these cells to die off once the light organ is sufficiently colonized. The light organ of certain squid contains reflective plates that intensify and direct the light produced, due to proteins known as reflectins. They regulate the light to keep the squid from casting a shadow on moonlit nights, for example. Sepolid squids expel 90% of the symbiotic bacteria in its light organ each morning in process known as "venting." Venting is hypothesized to provide the free-living inoculum source for newly hatched squids [48(b), 50, 51].

The specific symbionts in the light-emitting organs of certain squids and fishes, where it produces luminescence by expressing the *lux operon*, a small cluster of genes found in several of the Vibrionaceae. Acylhomoserine lactone quorum sensing controls luminescence, which was discovered in *V. fischeri* but is a common feature of host-associated bacteria in a number of genera. This particular nonpathogenic member of the Vibrionaceae is an ideal candidate for comparative genome analyses with pathogenic vibrios. The best understood of the *V. fischeri* symbiotic associations are that with sepiolid squids. These symbioses involve monospecific populations of *V. fischer* *cultured* extracellularly, but within epitheliumlined crypts, in a specialized host organ. The squid associations have been extensively studied because of the ease of initiating the association and of observing developmental changes in both partners. Nevertheless, important questions remain concerning the genetic and metabolic mechanisms by which *V. fischeri* and other symbiotic bacteria adjust to the special environment of host tissue [52].

About toxicity data $Log[1/EC_{50}]$ of compounds of Figure 3.1 on *Vibrio fischeri* were considered from literature [53] through the work of Cronin [10]. They were determined in a static assay using the luminescent bacterium *Vibrio fischeri*. The endpoint measured is the concentration causing a 50% decrease in bacterial bioluminescence as compared to a control. Toxicity data from 15 min assay were preferred. If a compound had more than one measured toxicity value, a mean value was calculated. When 15 min data were not available, but the compound was tested in a 5 or 30 min assay, a toxicity value from a different time endpoint was used [10].

Worth noting that the bioluminescence bacteria test, Microtox, is an *in vitro* test system that uses bioluminescent bacteria for toxin detection from water, air, soil, and sediments. It is a metabolic inhibition test and encompasses acute toxicity and genotoxicity analysis. Microtox is using *Vibrio fischeri*, as a marine, non-pathogenic bacterium, employing its main property, that is, the giving light like a product of cellular breathing. The breathing is fundamental in cellular metabolism and in all life-associated processes. In this way, every inhibition of the cellular activity results in a decay of the respiration velocity and a resulting decrease of the luminescent power. The principle of those tests is that the non inhibited bacteria beam light, and those that are exposed to toxic substances, give less light, correlated with the inhibition degree. Those measurements of the light emission are done to sometime intervals, usually after 5 and 15 min, with the aid of some instruments that measured the light emission. The test is performed by the adding of photoluminescence bacteria in different dilutions of the test sample. The samples are done to 15°C for assuring the vitality of bacteria [54].

Pimephales promelas

Classified as: Kingdom: *Animalia* > Phylum: *Chordata* (animals with a spinal chord) > Subphylum: *Vertebrata* (animals with a backbone) > Superclass: *Osteichthyes* (bony fishes) > Class: *Actinopterygii* (ray-finned and spiny rayed fishes) > Subclass: *Neopterygii* > Infraclass: *Teleostei* > Superorder: *Ostariophysi* > Order: *Cypriniformes* (minnows and suckers) > Family: *Cyprinidae* (carps and minnows) > Genus: *Pimephales* (the bluntnose minnows), *Pimephales promelas* or Fathead minnow (in English coming from to the enlarged head of breeding males) [55(a)] are usually found in Midwest and Great Plains states, through the Great Lakes basin to New York, south to Texas and New Mexico, and north into the Yukon, as well as, in all drainage in Minnesota [55(b)], see the Figure 3.4.

It is the most common species of minnow in the state. They live in many kinds of lakes and streams, but are especially common in shallow, weedy lakes, bog ponds; low-gradient, turbid (cloudy) streams; and ditches [55(a)]. Often found in large schools around submerged structure. These habitats often have very tolerant of muddy water, low oxygen, and a wide range of pH levels [55 (a) and (b)] and no predators [55(b)].

Figure 3.4. An exemplar of the Fathead minnow or Pimephales promelas species [55 (c)].

Sexual maturity of *P. promelas* is reached at one year of age. Rapid rate of reproduction makes it an excellent choice for stocking where predatory fish are present. Spawns from early May through August, when water temperature exceeds 16°C. The male selects the nest site, which normally is under an object such as a log, rock, stick, pop can, or whatever may be dumped at the bottom of the waterway. Females produce clutches of eggs (groups of eggs that become ready for spawning at the same time). Each clutch may contain 80–370 eggs. Most females probably spawn several clutches in a season. The adhesive eggs are deposited on the under surface of floating objects, and the male guards them. The eggs (embryos) hatch in 4–6 days [55(a) and (b)].

Fatheads only grow to about 65–70 mm and males grow bigger than females [55(a)]. Most of these little fish live for only one year. Less than 20% of one-year-olds live to two years old. On rare occasions a fathead makes it to three years old [55(b)]. Nevertheless, Fatheads are noted for their ability to withstand low oxygen levels [55(a)], although, they commonly occur with white suckers (*Catostomus commersoni*), common shiners (*Luxilus cornutus*), northern redbelly dace (*Phoxinus eos*), creek chubs (*Semolitus atromaculatus*), black bullheads (*Ameiurus melas*) [55(a)]. Fathead minnows are considered an opportunist feeder.

They eat just about anything that they come across, such as algae, protozoa (like ameba), plant matter, insects (adults and larvae), rotifers, and copepods [55(a)], as well as primarily zooplankton (microscopic crustaceans), microscopic plants, and occasionally other fishes [55(b)].

Instead, in lakes and deeper streams, fatheads are common prey for: crappies (*Pomoxis nigromaculatus*), rock bass, perch walleyes (*Perca flavecens*), largemouth bass (*Micropterus salmoides*), northern pike (*Esox lucius*), snapping turtles, herons, kingfishers, and terns. Eggs of the fathead are eaten by: painted turtles and certain large leeches [55(a)]. Concerning the interaction with human, although humans do not eat fatheads, they harvest them as bait since they are reared in ponds for the bait industry.

Fathead minnows are probably the most abundant minnow in Minnesota, and so they have no special conservation status, being the premier bait minnow in Minnesota [55(a)].

Finally, acute toxicity data $Log(1/EC_{50})$ of chemical compounds of Figure 3.1 to *Pimephales promelas* were considered from the measurements and according with the protocol described by Russom et al. [56] from the paper of Cronin and co-workers, who did the effort to bring up together a lot of trusting experimental data [10].

MODELING THE ECOTOXICO-LOGISTICAL BATTERIES

The Concise SPECTRAL-SAR Algorithm

Without enter into details [57, 58], if one has to solve correlation between a set of biological activities of N-compounds with the set of M-structural properties of each of them, a spectral algebraic algorithm can apply as follows:

First, the N-biological activities, as well as their respective predictor variables are grouped in vectors, as displayed in Table 3.2, to which the additional unity vector $|X_0\rangle = |1 \ 1 \ ... \ 1\rangle$ was added.

Table 3.2. Synopsis of the SPECTRAL-SAR descriptors.

Activity	Structural Predictor Variables					
$|Y\rangle$	$|X_0\rangle$	$|X_1\rangle$	$\cdots$	$|X_M\rangle$	$\cdots$	$|X_M\rangle$
y_1	1	x_{11}	$\cdots$	x_{1k}	$\cdots$	x_{1M}
y_2	1	x_{21}	$\cdots$	x_{2k}	$\cdots$	x_{2M}
$\vdots$	$\vdots$	$\vdots$	$\vdots$	$\vdots$	$\vdots$	$\vdots$
y_N	1	x_{N1}	$\cdots$	x_{Nk}	$\cdots$	x_{NM}

Then, the searched correlation equation takes the spectral decomposition of the activity vector $|Y\rangle$ on the vectorial base $\{|X_0\rangle, |X_1\rangle, ..., |X_k\rangle, ..., |X_M\rangle\}$ with the form:

$$|Y\rangle = B_0|X_0\rangle + B_1|X_1\rangle + ... + B_k|X_k\rangle + ... + B_M|X_M\rangle + |e\rangle. \qquad (3.7)$$

The general equation (3.7) is unpractical since the appearance of the error vector $|e\rangle$ [58]. However, by an algebraic optimization method it can be smeared out leaving with the explicitly unfolded determinant equation

$$\begin{vmatrix} |Y\rangle & \omega_0 & \omega_1 & \cdots & \omega_k & \cdots & \omega_M \\ |X_0\rangle & 1 & 0 & \cdots & 0 & \cdots & 0 \\ |X_1\rangle & r_0^1 & 1 & \cdots & 0 & \cdots & 0 \\ \vdots & \vdots & \vdots & \vdots & & \vdots \\ |X_k\rangle & r_0^k & r_1^k & \cdots & 1 & \cdots & 0 \\ \vdots & \vdots & \vdots & \vdots & & \vdots \\ |X_M\rangle & r_0^M & r_1^M & \cdots & r_k^M & \cdots & 1 \end{vmatrix} = 0, \qquad (3.8)$$

once solved respecting $|Y\rangle$ [58], where the involved parameters are given with the scalar product formulas:

$$r_i^k = \frac{\langle X_k | \Omega_i \rangle}{\langle \Omega_i | \Omega_i \rangle}, \, \omega_k = \frac{\langle \Omega_k | Y \rangle}{\langle \Omega_k | \Omega_k \rangle}, \, k = \overline{0, M}, \qquad (3.9)$$

while the orthogonal basis $\left\{ |\Omega_0\rangle, |\Omega_1\rangle, ..., |\Omega_k\rangle, ..., |\Omega_M\rangle \right\}$ was introduced according with the Gram-Schmidt rules:

$$|\Omega_0\rangle = |X_0\rangle, |\Omega_k\rangle = |X_k\rangle - \sum_{i=0}^{k-1} r_i^k |\Omega_i\rangle, \, k = \overline{1, M} \qquad (3.10)$$

without concerning on the order of orthogonalization of the structural and activity vectors of Table 3.2.

From this algorithm there is clear that no statistical tool was invoked hitherto thus furnishing in principle an exact unequivocal structure-activity equation with a high predictable character.

Such a methodology was already applied to study the enzymatic activity in a particular case of acetylcholinesterase [57(b)], as well as predicting the hierarchies of the *Tetrahymena pyriformis'* ecotoxicology [58] with quite impressive results. It will be thus considered also in the following.

The LOGISTIC-SPECTRAL-SAR Computational Strategy

Basically, the previously described SPECTRAL-SAR procedure is circumvented to the consecrated Hansch quantitative structure-activity relationships (QSARs) that prescribe the activity expansion under the generic minimal yet meaningful form:

$$A = B_0 + B_1 \left(\begin{array}{c} electronic \\ parameter \end{array} \right) + B_2 \left(\begin{array}{c} hydrophobic \\ parameter \end{array} \right) + B_3 \left(\begin{array}{c} steric \\ parameter \end{array} \right) \qquad (3.11)$$

Nevertheless, beside the already traditional regression methods in determining the correlation coefficients of the expression (3.11) [59], the most recent SPECTRAL-SAR orthogonal method was found with many practically advantages:

- it gives a similar prediction respecting the traditional quantitative structure-activity regression analysis so being capable to replace it;
- it has the strength of no dependency on the way in which the input data are considered thus being lest dependent on the outliers detection being also more direct [57];
- it uses the algebraically instead of statistically recipe so furnishing a substitute of estimated coefficients of structure-activity correlation;
- it is easily applicable also to the case in which the number of structural parameters exceeds those of the available biological activities, a situation more often meet in actual practice but being still an open problem in QSAR due to the statistical forbidden condition that such situations imply [58];

- it is able to furnish also the key in treating the so called spectral analysis of the activity itself through *action norm* $\|Y\rangle\|$ and its *least activity path principle* (i.e., $\delta\|Y\rangle\| = 0$ over many possible end-points) thus providing the appropriate mechanistic picture of the envisaged system [58].

In this study, the biologically activities were took from the recent work of Cronin and collaborators [10], which select diverse chemical structures incorporating narcosis, as well as other more specific mechanism of toxic action. Also the chemicals were required to span an enough range of hydrophobicity. Actually, the compounds of Figure 3.1 fully satisfy both dissimilarity- and similarity-based selection criteria.

However, the workable activities stand as the toxicity data $Log(1/EC_{50})$. In the case of the unicellular alga [10] (*Chlorella vulgaris*) they were collected in a biochemical assay (15 min) by the method which has been found to be an excellent predictor of the toxicity of chemicals to other species [46], so with good application in ecotoxicology.

On the other side, the chemical structures of the molecules of Figure 3.1 were computed by the HyperChem Software (7 Release version) [60] providing that their 3D structures were optimized by AM1 semiempirical calculation while molecular dynamics involved the Polak-Rebier algorithm to reach to 0.01 root mean square gradient. The results are as well displayed in Table 3.3 as *LogP*, *POL*, and E_{TOT} for the hydrophobic, electronic, and steric parameters of equation (3.11), respectively.

Table 3.3. Hydrophobic LogP, electronic POL, and optimized (steric) total energy E_{TOT} parameters, computed within HyperChem 7.0 environment [60], for the series of compounds of Figure 3.1 tested upon *Chlorella vulgaris*, *Vibrio fischeri*, and *Pimephales promelas* species providing experimental $Log(1/EC_{50})$ activities, A$_1$, A$_2$, and A$_3$, respectively [10].

No.	Parameters Compd.	A$_1$	A$_2$	A$_3$	1	LogP	POL	E$_{TOT}$
		$\|Y_1\rangle$	$\|Y_2\rangle$	$\|Y_3\rangle$	$\|X_0\rangle$	$\|X_1\rangle$	$\|X_3\rangle$	$\|X_3\rangle$
1	CH$_3$OH	−4.06	−3.21	−2.96	1	−0.27	3.25	−11622.9
2	C$_2$H$_5$OH	−3.32	−2.7	−2.49	1	0.08	5.08	−15215.4
3	C$_4$H$_9$OH	−2.73	−1.64	−1.37	1	0.94	8.75	−22402.8
4	C$_4$H$_8$O	−2.51	−1.76	−1.65	1	1.01	8.2	−21751.8
5	C$_5$H$_{10}$O	−2.23	−0.99	−1.25	1	1.64	10.04	−25344.6
6	C$_6$H$_5$OH	−1.46	0.5	0.51	1	1.76	11.07	−27003.1
7	C$_6$H$_5$NH$_2$	−1.34	−0.3	−0.16	1	1.26	11.79	−24705.9
8	CH$_3$-C$_6$H$_4$-OH	−1.01	1.16	0.29	1	2.23	12.91	−30597.6
9	OH-C$_6$H$_4$-O-CH$_3$	−0.97	1.46	0.05	1	1.51	13.54	−37976.3
10	OH-C$_6$H$_4$-NH$_2$	−0.91	−0.28	1.65	1	0.98	12.42	−32095.4
11	C$_6$H$_5$-CHO	−0.81	1.32	1.14	1	1.72	12.36	−29946.9
12	CH$_3$-C$_6$H$_4$-OH	−0.81	0.74	0.89	1	2.23	12.91	−30597.2
13	C$_6$H$_3$(CH$_3$)$_2$OH	−0.65	2.4	0.94	1	2.7	14.74	−34190.8

Table 3.3. *(Continued)*

No.	Parameters Compd.	A₁ $\lvert Y_1 \rangle$	A₂ $\lvert Y_2 \rangle$	A₃ $\lvert Y_3 \rangle$	1 $\lvert X_0 \rangle$	LogP $\lvert X_1 \rangle$	POL $\lvert X_3 \rangle$	E_{TOT} $\lvert X_3 \rangle$
14	CH_3-C_6H_4-NO_2	−0.5	1.54	0.73	1	0.94	13.98	−42365.1
15	C_6H_5-O-Cl	−0.42	1.15	1.32	1	2.28	13	−35307.6
16	$C_6H_3(NO_2)NH_2$	−0.36	0.56	1.07	1	−1.75	15.22	−63030.2
17	$C_{11}H_8O_2$	0.16	2.81	3.19	1	2.39	20.99	−49768.3
18	$C_6H_4Cl_2$	0.37	1.56	1.19	1	3.08	14.29	−36217.2
19	$C_6H_3(NO_2)OH$	0.4	1.28	1.14	1	1.67	14.5	−65318
20	$C_6H_4N_2O_4$	0.41	3.07	2.37	1	1.95	13.86	−57926.7
21	$C_7H_6(NO_2)_2$	0.7	0.55	0.87	1	2.42	15.7	−61520.7
22	$C_{15}H_{23}OH$	1.45	1.41	2.78	1	5.48	27.59	−59316.5
23	$C_6H_3NCl_4$	1.48	2.15	2.93	1	3.34	19.5	−57920.2
24	C_6Cl_5OH	1.69	2.45	3.08	1	−0.54	20.71	−68512.4
25	$C_{12}H_{10}N_2O$	2.16	2.41	2.23	1	4.06	22.79	−55488.9
26	C_6Br_5OH	3.1	2.74	3.72	1	5.72	24.2	−66151.5

The employed toxicity data for the chemical set of compounds of Figure 3.1 on *Chlorella vulgaris*, *Vibrio fischeri*, and *Pimephales promelas* species are presented in the Table 3.3 and meet most of the criteria for high quality data, that is, it has been produced to a standard protocol, in a single laboratory, by a single worker. These toxicological data have previously been evaluated (and undergone process of pre-validation) by the development of QSARs for nonpolar and polar narcosis [46] and investigated by QSARs with other species such as *T. pyriformis* [22, 61]. However, the acute toxicity, assessed in short and low-cost unicellular tests, is also considered to be a surrogate for the prediction of toxicity to higher aquatic organisms [62].

Nevertheless, for completing the *logistic-spectral analysis* the next steps are assumed:

i.   for the activities of Table 3.3, their vectorial form is achieved through applying of the SPECTRAL-SAR algorithm:

$$\lvert Y_i \rangle^{ENDPOINT} = B_0 \lvert X_0 \rangle + B_1 \lvert X_1 \rangle + B_2 \lvert X_2 \rangle + B_3 \lvert X_3 \rangle, \ i = \overline{1,3} ; \qquad (3.12)$$

the predicted spectral norm can be draw, $\lVert Y_i \rVert$, $i = \overline{1,3}$, for each envisaged species;

ii. the initial chemical concentration of logistical chemical–biological progress curve equation (3.6) is identified with the predicted SPECTRAL-SAR activity norms:

$$[\chi_{0(i)}] \rightarrow \left\lVert Y_i \right\rangle^{ENDPOINT} \right\rVert, i = \overline{1,3} ; \qquad (3.13)$$

based on idea that the evolution of the chemical concentration producing a biological effect starts evolving from the predicted (computed) activity and diminished in time under the environmental and bio-degrability effects;

iii. in the same heuristically line the real maximum biological effect in chemical–biological equation (3.6) would be seen as the positive reminiscence of the predicted SPECTRAL-SAR activity against the measured activity:

$$\beta_{max(i)} \rightarrow \sqrt{\left(\left\|\langle Y_i \rangle^{ENDPOINT}\right\| - \left\|\langle Y_i \rangle^{EXP}\right\|\right)^2}, i = \overline{1,3};$$ (3.14)

iv. in these conditions, the computational EC_{50} parameter of chemical–biological equation (3.6) is as well considered as the positive reminiscence of the predicted SPECTRAL-SAR activity against its average activity:

$$EC_{50(i)} \rightarrow \sqrt{\left(\left\|\langle Y_i \rangle\right\| - \overline{|Y_i\rangle}\right)^2}, i = \overline{1,3};$$ (3.15)

v. the progress curves of the so constructed biological–chemical activities (3.6) are computed and from them the cut-off time extracted:

$$\tau_{\infty(i)} = \lim_{\beta_{(i)} \rightarrow 0}\left[1 - \frac{1}{\ln(t+e)}\right], i = \overline{1,3}$$ (3.16)

for each envisaged species in ecotoxicological battery;

vi. the obtained times are grouped, for each concerned species, and the inter-species mechanistic hierarchies are constructed according with the *least-action activity principle* of predicted norm-cut-off time paths

$$\left[A_{(i)}, B_{(i)}\right] = \sqrt{\left(\left\|Y^B_{(i)}\right\| - \left\|Y^A_{(i)}\right\|\right)^2 + \left(\ddot{A}^B_{\infty(i)} - \ddot{A}^A_{\infty(i)}\right)^2}, i = \overline{1,3}; A,B{:}ENDPOINTS$$ (3.17)

providing that *the shortest path the faster biological–chemical interaction is activated, thus assuring the more effective (toxico) chemical effect on biological system.*

The practical application of the current spectral-logistical algorithm is in next exposed and interpreted for the chemical–biological interactions of the biological species of previous section with the chemical compounds of Figure 3.1 by employing the data of Table 3.3.

The Logistic-spectral SAR Ecotoxicological Analysis

The previously presented SPECTRAL-SAR-logistical algorithm is now applied upon the data of Table 3.3 providing, in the first stage the results reported in Tables 3.4–3.6 for *Chlorella vulgaris, Vibrio fischeri,* and *Pimephales promelas* species, respectively.

Table 3.4. Spectral structure-activity relationships (SPECTRAL-SAR) predicted with all possible correlation models (end points) considered from data of Table 3.3, together with the experimental or measured activity of *Chlorella vulgaris* species, paralleling the corresponding correlation factor, spectral norm (3.13), and the asymptotic cut-off time (3.16) of the associated logistical-spectral model through equations (3.6) and (3.13)–(3.16).

Activity (SPECTRAL-SAR Equation) Model (Endpoint)	r	$\lVert /Y_I \rangle \rVert$	τ_∞
$\lvert Y_1 \rangle^{Ia} = -1.65895 \lvert X_0 \rangle + 0.634092 \lvert X_1 \rangle$	0.618968	5.76725	0.436918
$\lvert Y_1 \rangle^{Ib} = -4.48035 \lvert X_0 \rangle + 0.279384 \lvert X_2 \rangle$	0.923589	8.18957	0.440029
$\lvert Y_1 \rangle^{Ic} = -4.00411 \lvert X_0 \rangle - 0.0000865456 \lvert X_3 \rangle$	0.887337	7.89593	0.439784
$\lvert Y_1 \rangle^{IIa} = -4.46015 \lvert X_0 \rangle + 0.0222605 \lvert X_1 \rangle + 0.275066 \lvert X_2 \rangle$	0.923735	8.19075	0.44003
$\lvert Y_1 \rangle^{IIb} = -4.24209 \lvert X_0 \rangle + 0.380779 \lvert X_1 \rangle - 0.0000748672 \lvert X_3 \rangle$	0.954562	8.44123	0.440222
$\lvert Y_1 \rangle^{IIc} = -4.65267 \lvert X_0 \rangle + 0.180771 \lvert X_2 \rangle - 0.0000388795 \, X_3 \rangle$	0.951655	8.41758	0.440204
$\lvert Y_1 \rangle^{III} = -4.50442 \lvert X_0 \rangle + 0.234126 \lvert X_1 \rangle + 0.0986645 \lvert X_2 \rangle$ $-0.00000533489 \lvert X_3 \rangle$	0.963357	8.51281	0.440274
$\lvert Y_1 \rangle^{EXP}$	1.000000	8.81159	0.440478

Table 3.5. Spectral structure-activity relationships (SPECTRAL-SAR) predicted with all possible correlation models (end points) considered from data of Table 3.3, together with the experimental or measured activity of *Vibrio fischeri* species, paralleling the corresponding correlation factor, spectral norm (3.13), and the asymptotic cut-off time (3.16) of the associated logistical-spectral model through equations (3.6) and (3.13)–(3.16).

Activity (SPECTRAL-SAR Equation) Model (Endpoint)	r	$\lVert Y_2 \rangle \rVert$	τ_∞
$\lvert Y_2 \rangle^{Ia} = -0.201262 \lvert X_0 \rangle + 0.52453 \lvert X_1 \rangle$	0.51196	5.90227	0.436419
$\lvert Y_2 \rangle^{Ib} = -2.64591 \lvert X_0 \rangle + 0.238822 \lvert X_2 \rangle$	0.789409	7.79875	0.440075
$\lvert Y_2 \rangle^{Ic} = -2.15959 \lvert X_0 \rangle - 0.0000720417 \lvert X_3 \rangle$	0.738546	7.43153	0.439601
$\lvert Y_2 \rangle^{IIa} = -2.65657 \lvert X_0 \rangle - 0.0117534 \lvert X_1 \rangle + 0.241101 \lvert X_2 \rangle$	0.789456	7.79909	0.440076
$\lvert Y_2 \rangle^{IIb} = -2.35539 \lvert X_0 \rangle + 0.313287 \lvert X_1 \rangle - 0.0000624332 \lvert X_3 \rangle$	0.793251	7.82676	0.440109
$\lvert Y_2 \rangle^{IIc} = -2.76727 \lvert X_0 \rangle + 0.169375 \lvert X_2 \rangle - 0.0000273805 \lvert X_3 \rangle$	0.805768	7.91823	0.440214
$\lvert Y_2 \rangle^{III} = -2.68588 \lvert X_0 \rangle + 0.128529 \lvert X_1 \rangle + 0.124301 \lvert X_2 \rangle - 0.00000353238 \lvert X_3 \rangle$	0.809947	7.94886	0.440248
$\lvert Y_2 \rangle^{EXP}$	1.000000	9.37758	0.441463

Table 3.6. Spectral structure-activity relationships (SPECTRAL-SAR) predicted with all possible correlation models (end points) considered from data of Table 3.3, together with the experimental or measured activity of *Pimephales promelas* species, paralleling the corresponding correlation factor, spectral norm (3.13), and the asymptotic cut-off time (3.16) of the associated logistical-spectral model through equations (3.6) and (3.13)–(3.16).

Activity (SPECTRAL-SAR Equation) Model (Endpoint)	r	$\lVert Y_3 \rangle \rVert$	τ_∞
$\lvert Y_3 \rangle^{Ia} = -0.209645 \lvert X_0 \rangle + 0.566471 \lvert X_1 \rangle$	0.53973	6.4009	0.437342
$\lvert Y_3 \rangle^{Ib} = -3.12624 \lvert X_0 \rangle + 0.277169 \lvert X_2 \rangle$	0.894346	8.90932	0.440991
$\lvert Y_3 \rangle^{Ic} = -2.47683 \lvert X_0 \rangle - 0.0000815288 \lvert X_3 \rangle$	0.815901	8.32132	0.440454

Table 3.6. *(Continued)*

Activity (SPECTRAL-SAR Equation) Model (Endpoint)	r	$\|\mid Y_3\!> \mid\|$	τ_∞
$\mid Y_3\!>^{IIa} = -3.20607 \mid X_0\!> - 0.0880001 \mid X_1\!> + 0.294236 \mid X_2\!>$	0.896578	8.92624	0.441005
$\mid Y_3\!>^{IIb} = -2.67949 \mid X_0\!> + 0.324268 \mid X_1\!> - 0.0000715835 \mid X_3\!>$	0.866744	8.70086	0.440815
$\mid Y_3\!>^{IIc} = -3.23924 \mid X_0\!> + 0.212504 \mid X_2\!> - 0.0000254953 \mid X_3\!>$	0.90632	9.00023	0.441064
$\mid Y_3\!>^{III} = -3.22812 \mid X_0\!> + 0.0175597 \mid X_1\!> + 0.206346 \mid X_2\!> - 0.0000265806 \mid X_3\!>$	0.906386	9.00074	0.441065
$\mid Y_3\!>^{EXP}$	1.000000	9.72062	0.441561

At first glance, from Tables 3.4–3.6, there is clear that the correlation factor parallels the spectral norm and temporal cut-off limits, in each reported case, including the experimental or measured situation. That is, as correlation factor increase as both spectral norm and cut-off time increase as well.

Nevertheless, when thinking to a mechanistic interpretation of the computed models and of the associate endpoints the least, or minimum, action principle within the *spectral norm-cut-off time* abstract space has to be consider. That would require that shortest distance between two endpoints' spectral norm and cut-off times implies the first chemical–biological interaction. Therefore, caution has to be paid to not confound the maximum-correlation factor-maximum-spectral norm-and-cut-off time behavior with the minimum distance between such spectral norm-cut-off time "points" in the abstract endpoint space.

To further emphasize on the minimum path principle, in Figure 3.5 there are displayed the decay on chemical–biological interactions, according with equation (3.6) and the replacement rules (3.13)–(3.15), for all considered endpoints (SPECTRAL-SAR predicted models along the experimental activity) for all species of Table 3.3, involved in the present ecotoxicological battery. It follows that since the shapes of all considered species-end-points display the same features their difference should be evidenced only through their extremes, that is, the initial chemical concentration (dose) and by their effectiveness in biological uptake (the time of action).

This way, through correspondence (3.13) and by equation (3.16), in fact, the main roles in establishing the endpoints parameters, when compared with others, refer to the spectral norm and cut-off time, respectively.

From these considerations follows the necessity of considering the spectral-norm-cut-off-time path (3.17) combination when one likes to asses through a single number (i.e., by a quantification procedure) the abstract, yet mechanistically, distance between two considered endpoints. However, even from representations of Figure 3.5 there comes out that *Pimephales promelas* species is characterized by the highest and most intensive, as well as the most extensive chemical–biological interaction respecting the other two species in battery. There is, therefore, already clear that in an environmental context in which all three species exists and are affected by the same set of chemical compounds of Table 3.3 the *Pimephales promelas'* biological uptake comes firstly into act, that is, will be the first one affected.

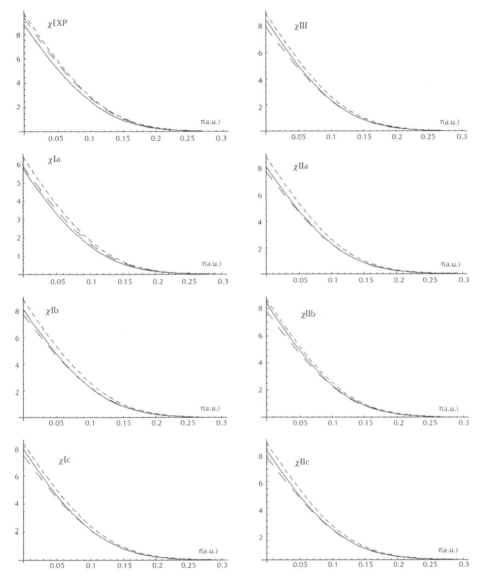

Figure 3.5. Temporal logistic-spectral representations of the *Chlorella vulgaris* (full line), *Vibrio fischeri* (interrupted line), and *Pimephales promelas* (dashed line) species for all endpoints of Tables 3.4–3.6 based on the equation (3.6) with the algorithm (3.13)–(3.15).

Nevertheless, to see the detailed mechanistically graph of dynamical actions and interactions intra- and inter-species of the ecotoxicological battery the least principle of chemical–biological interaction will be employed within the spectral norm-cut-off time with the help of the present SPECTRAL-SAR-logistical data of Table 3.7.

Table 3.7. Synopsis of the values of spectral norm-cut-off time paths (3.17) connecting the endpoints of SPECTRAL-SAR models of Tables 3.4–3.6 within the spectral norm-cut-off time abstract space ($\|Y_i>\|$ vs. $\tau_{\infty(i)}$) for the species of Table 3.3.

Paths among Endpoints	Values for Species		
	Chlorella v.	Vibrio f.	Pimephales p.
Ia-IIa-III	2.74556	2.04659	2.59984
Ia-IIb-III	2.74556	2.04659	2.59984
Ia-IIc-III	2.74556	2.04659	2.59984
Ib-IIa-III	0.323242	0.15011	0.0914192
Ib-IIb-III	0.323242	0.15011	0.508336
Ib-IIc-III	0.323242	0.15011	0.0914192
Ic-IIa-III	0.616885	0.51733	0.679419
Ic-IIb-III	0.616885	0.51733	0.679419
Ic-IIc-III	0.616885	0.51733	0.679419

The construction of such diagrams follows some preliminary or formal steps. Firstly, there is chosen that the species' endpoint states are represented by the height of the associated spectral norm (the equally choice would be performed in terms of cut-off time since parallels the spectral norm in all cases).

Then a grid of the endpoints is considered according with the general tendency from minimum to maximum paths of Table 3.7 such that once a model or endpoint is nominated across a path it will be not repeated across other.

For instance, identifying that the shortest path across all endpoints, when from each class of SAR models a single model is picked up or touched by chemical–biological evolution, one gets that the paths *Ib-IIa-III* and *Ib-IIc-III* are the two possible candidates. Further chose will be made in the light of the so called *local* minimum path principle: the *Ib-IIa-III* path will be preferred since the state *Ib* is closer to *IIa* rather to the *IIc* one for all species of Tables 3.4–3.6.

Thus, the chart will start with *Ib* followed by *IIa* endpoint regions across ecotoxicological battery. Next, we go to the next group of path from where the routes *Ic-IIa-III*, *Ic-IIb-III*, and *Ic-IIc-III* are the candidates.

The route *Ic-IIa-III* is now out of discussion since we already accommodated *IIa* after *Ib* on the planned ecotoxicological chart. Thus, there remains to chose between *Ic* followed by *IIb* or by *IIc*; analyzing the data of Tables 3.4–3.6 for all species we found out that from three concerned species for two of them, namely for *Vibrio fischeri* and *Pimephales promelas*, the spectral norms of *IIb* endpoints rather that those of *IIc* situates closer to *Ic*. As such we conclude that the next order will be *Ic* followed by *IIb*.

At the end from the third group of paths in Table 3.7, it follows that *Ia* can only be followed by the remaining *IIc* set of endpoints since all other were previously accommodated in the chart.

Worth noted that this chart ordering was identically with that considered in previous study of *Tetrahymena pyriformis* in relation with the same set of chemical compounds [58]. However, with all these considerations the spectral chart of Figure 3.6 can be drawn.

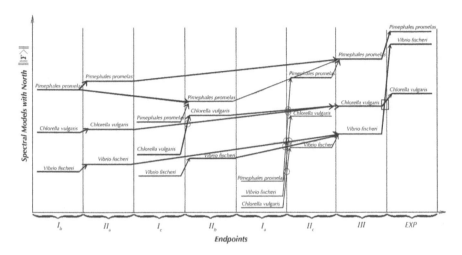

Figure 3.6. The spectral representation of the chemical–biological interaction paths across the SPECTRAL-SAR to experimental endpoints for the *Chlorella vulgaris, Vibrio fischeri,* and *Pimephales promelas* species, according with least path rule within the spectral norm-cut-off time space applied on Table 3.7 results, respectively. The primary, secondary, and tertiary least path hierarchies of Table 3.7 are represented by decreasing the thickness of the connecting lines, while the interferences between species interactions is indicated by a rectangular mark for primary hierarchy and by circle marks for secondary and tertiary hierarchies.

In Figure 3.6 appears that each endpoint region contains all *Chlorella vulgaris, Vibrio fischeri,* and *Pimephales promelas* species with their relative spectral norms. Now, they will be all connected through paths according with minimum path principles applied intra-species this time for the paths of Table 3.7. In this respect, it follows that the *Pimephales promelas* species provides the first, the most preeminent path, as already anticipated from Figure 3.5, from the one-parameter model (*Ib* or *POL*arizability based model) touching *IIa* one (based on electronic-*POL* followed by hydrophobicity-*Log* model) until the measured endpoint (i.e., the experimental one), not before that the steric effects (comprised in the *III*) to enter ultimately in action. Similar mechanistic analyses can be performed for all other connected endpoints.

However, there is clear from Figure 3.6 that three major hierarchies can be formulated: the primary one that starts from *Ib* (*POL*), the secondary one that starts from *Ic* (E_{TOT}), and the tertiary one that starts from *Ia* (*Log*) SAR models for all three species, respectively. The difference between these three hierarchies stands in their efficacies in action so that the longer path the more subtle (or less intensive) chemical–biological interaction associates with. Still, these subtle interaction becomes important in end-point-chart regions where their register interferences with other superior hierarchies, as is the case in Figure 3.6 also.

The inter-path crossings are marked in Figure 3.6 with a rectangular mark for primary hierarchy and by circles for the rest. The interference between the paths of primary hierarchy appears in region *III* between *Chlorella vulgaris* and *Vibrio fischeri* species only. That means that since *Pimephales promelas* species is firstly affected (at-tacked) by the set of compounds of Table 3.3 the other two species interchange their chemical–biological effects only after all hydrophobic, electronic, and steric effects are consumed. In other words, primarily, the effects on *Pimephales promelas* does not affect the other two species while the effects on one of the *Chlorella vulgaris* and *Vibrio fischeri* species induce (or transfer) similar toxic effects on the other. Moreover, in this stage, the *Pimephales promelas* species is once more isolated from the rest of species through additional path *Ib-IIb-III* that do not interfere with other species' paths.

Remarkably, when about the secondary hierarchies the only interference is recod-ed at *Chlorella vulgaris* between its primary and the secondary *toxicological waves*, associated with the second generation of the longer spectral-norm-cut-off time paths. Therefore, this second stage of ecotoxicological action tell us that the *Chlorella vulgaris* will be further affected, however not transferring the effects upon it to the rest of species in the ecosystem. Moreover, this reinforcing toxicological effect takes places from *Ic* to *IIb* endpoint regions in accordance with the fact that the combination of hydrophobicity with doubled steric effects.

Finally, the third stage, or wave of toxicological effects (the endpoints) suggest that even being associated with the less effective chemical–biological action the ter-tiary hierarchy becomes important by means of the numerous interference that pro-duce with both primary and secondary effects. For instance, the tertiary effects on the *Chlorella vulgaris* jointly affects the primary and secondary effects on *Vibrio fischeri* species, thus, sustaining in this more subtle way their ecotoxicological activity. More-over, is in this third stage that, for the first time, the path of *Pimephales promelas* species interferes with those of *Chlorella vulgaris* and *Vibrio fischeri* species in both their secondary and primary stages of toxically action. This behavior stands also as a computational confirmation of the empirical rule according with, indeed, in an eco-system once one species is affected all other existing ones are soon or later, directly or by interference of effects, affected as well. Nevertheless, the present SPECTRAL-SAR-logistical model combined with the least principle of action gives a detailed map of these inter-actions both at the inter- and intra-species level. At the same time, the chart of Figure 3.6 has the important advantage that suggest by its mechanistically in-terpretation at which level and upon which structural property (Log, POL or E_{TOT}) one can act in order to promote or to inhibit a certain action through co-existing species. It furnishes, this way, in fact, a model battery of ecotoxicological action, here, in its spectral-ecotoxico-logistical version.

CONCLUSION

The EU White Paper policy on chemical (from 2001 ahead) wants to identify hazards of all chemicals on the market greater than one tone [63]. In this respect, (Q)SARs methods have been found most useful for identifying these hazards to limit animal testing, since the current policy for the risk evaluation of chemicals is to involve more (Q)SARs rather than animals testing [64]. Nevertheless, in 2003 the European Commission (EC) adopted a legislative proposal for a new chemical management system called REACH (Registration, Evaluation and Authorization of Chemicals) according which, for instance, all chemicals marketed in Europe in volumes greater than one tone per year should be granted within 11 years of entry into force, while physico-chemical and toxicity data for High Production Volume (HPV) chemicals should be available within 7 years [65, 66].

The main idea behind these global regulations is to use rapid, low cost computational models prior going to the experimental testing. However, most of the published models until now lack details of their statistical validation and definitions of the model domains of applicability [67]. So, such models do not fully meet the OECD (Organization for Economic Cooperation and Development) principles for QSAR validations [68], which is likely to interrupt their use for regulatory purposes [65]. In this respect, Huzelbos and Posthumus concluded that the current risk assessment procedures for chemicals in EU are time consuming and complex processes on the reason that the minimum data requirements for performing risk evaluation for high production volume chemicals appeared to be insufficient. The entire risk assessment to human and environment is carried out only for some well-selected substances. The resonance between industry and several EU member states involve very time consuming processes as well, although (Q)SARs and *in vitro* testing have already been advanced as alternatives for animal testing [37].

With these there is clear that there is an increasing interest in the using of toxicological-based quantitative structure-activity relationships as non-animal methods to provide data for priority setting, risk assessment, and chemical classification and labeling [19, 69]. The interspecies QAA(activity-activity)Rs are useful because they can be exploited to predict unknown toxicity and to verify the validity of other toxicity tests, quoting that strong trend of increasing toxicity with increasing hydrophobicity underpins such data sets [10]. However, QSARs strategies are increasingly being used as a tool to assist regulatory agencies in toxicological assessment of chemical and require appropriate validation at all stages of development [70]. When assessing to model ecotoxicological batteries reliable toxicity data are required for all trophic levels of the environment to ensure appropriate risk assessment of chemicals. With all these, there are currently insufficient published toxicity data and models to meet the needs of regulation and QSAR mechanistically interpretation [10, 21]. Usually, in the case of certain endpoints, the biological variability may be too large to enable reasonable quantitative predictions to be made, so the modeler may decide to convert the data into one or more categories of toxic effect [71].

In this context, the present work firstly proposes the general logistical equation for the chemical–biological interactions, further combined with the recent SPECTRAL-

SAR computational QSAR method [57, 58], and then jointly applied on the *Chlorella vulgaris*, *Vibrio fischeri*, and *Pimephales promelas* species viewed as an ecotoxicological battery. The main found addresses the synergetic behavior of the interspecies interference across the endpoints effects; that is there was analytically proved that the lower chemical–biological interactions may influence the main toxicological pathways through a sort of feed-back effect thus assuring the unitary modeling of ecotoxicological battery. The present SPECTRAL-SAR-Ecotoxico-Logistical (SPECTRAL-SAR-EL) analysis enlighten on the reliability of the least principle of paths across endpoints providing the most wanted mechanistic interpretation of the proposed model respecting the entry descriptors used.

APPENDIX: LOGISTIC ENZYME KINETICS

We use a probabilistic approach, based on the law of mass action, to characterize *in vitro* enzymatic reactions of E-S type of Table 3.1:

$$1 = P_{REACT}([S]_{bind}) + P_{UNREACT}([S]_{bind}) . \quad (3.A1)$$

In equation (3.A1), $P_{REACT}([S]_{bind})$ is the probability that the enzymatic reaction of Table 3.1 proceeds at a certain concentration of substrate binding to the enzyme $[S]_{bind}$. The limits are:

$$P_{REACT}([S]_{bind}) = \begin{cases} 0 & , \quad [S]_{bind} \to 0 \\ 1 & , \quad [S]_{bind} >> 0 \end{cases} \quad (3.A2)$$

Note that $P_{REACT}([S]_{bind}) = 0$ when the enzymatic reaction does not proceed or when it stops because the substrate fails to bind or is entirely consumed.

Conversely, $P_{REACT}([S]_{bind}) = 1$ when the enzymatic reaction proceeds, and it is related to the standard quasi-steady-states approximation (QSSA). The probability of the occurrence of products in E-S reactions lies between these limits. Similarly, in the case where enzymatic catalysis does not take place, $P_{UNREACT}([S]_{bind})$, the limits are:

$$P_{UNREACT}([S]_{bind}) = \begin{cases} 1 & , \quad [S]_{bind} \to 0 \\ 0 & , \quad [S]_{bind} >> 0 \end{cases} \quad (3.A3)$$

This probabilistic treatment of enzymatic kinetics is based on the chemical bonding behavior of enzymes that act upon substrate molecules through diverse mechanisms and it may offer the key to the quantitative treatment of different types of enzyme catalysis.

To unfold the terms of equation (3.A1) to analyze E-S reactions we first recognize that the binding substrate concentration can be treated as the instantaneous substrate concentration: $[S]_{bind} = [S](t) .$

Maintaining the quasi-steady-state conditions for *in vitro* systems, we may assume constant association-dissociation rates so that probability of reaction is written as the rate of consumption of the substrate,

$$v(t) = -\frac{d}{dt}[S](t),$$

(3.A4)

to saturation:

$$P_{\text{REACT}}([S](t)) = \frac{v(t)}{V_{\text{max}}} = -\frac{1}{V_{\text{max}}}\frac{d}{dt}[S](t)$$

(3.A5)

after the initial transient of the enzyme-substrate adduct-complex interchanging.

We know only that expression (3.A5) behaves like a probability function, with values in the realm [0, 1]. Given expressions (3.A1), (3.A5), and the general Michaelis–Menten equation (see Appendix of the Chapter 2 of the present volume)

$$-\frac{d}{dt}[S](t) = \frac{V_{\text{max}}[S](t)}{[S](t) + K_M}$$

(3.A6a)

with the Michaelis constant

$$K_M = (k_{-1} + k_2)/k_1$$

(3.A6b)

we derive an expression for the unreacted probability term, $P_{\text{UNREACT}}([S](t))$. As such, the expression:

$$P_{\text{UNREACT}}([S](t))^{\text{MM}} = \frac{K_M}{[S](t) + K_M}$$

(3.A7)

satisfies all of the probability requirements, including the limits in (3.A3), and, when combined with equations (3.A1) and (3.A5), gives the instantaneous version of the classical Michaelis–Menten equation (A6). Remarkably, expression (3.A7) can be seeing as generalization of the efficiency of the Michaelis–Menten reaction under steady-state conditions. Originally, the efficiency depends on two parameters: K_M that embodies the thermodynamic conditions of the enzymic reaction and the initial substrate concentration $[S_0]$; it determines the ratio of the free to total enzyme concentration in the E-S reactions; that is, when the efficiency is equal to one, we cannot expect to find substrate free in the reaction, i.e. the E-S reactions are all consumed so that first branch of the limits (3.A3) is fulfilled as no further binding will occur.

It is clear that the Michaelis–Menten term (3.A6a) is just a particular choice for a probabilistic enzymatic kinetic model of the conservation law (3.A1). A more generalized version of equation (A6) that preserves all of the above probabilistic features may look like

$$P_{\text{UNREACT}}([S](t))^* = e^{-\frac{[S](t)}{K_M}}$$

(3.A8)

from which the Michaelis–Menten term (3.A6a) is returned by performing the $[S](t)$ first order expansion for the case where the bound substrate approaches zero:

$$P_{\text{UNREACT}}([S](t))^* = \frac{1}{e^{\frac{[S](t)}{K_M}}} \overset{[S](t) \to 0}{\cong} \frac{1}{1 + \frac{[S](t)}{K_M}} = P_{\text{UNREACT}}([S](t))^{\text{MM}}. \tag{3.A9}$$

Worth noting that there is no monotonically form between 0 and 1 other than that of equation (3.A8) to reproduce basic Michaelis–Menten term (3.A6a) when approximated for small $x = [S](t)/K_M$. For instance, if one decides to use $\exp(-x^2)$ then the unreactive probability will give $1/(1+x^2)$ as the approximation for small x, definitely different of what expected in basic Michaelis–Menten treatment (3.A6a). This way, the physico-chemical meaning of equation (3.A8) is that the Michaelis–Menten term (3.A7) and its associated kinetics apply to fast enzymatic reactions, that is, for fast consumption of $[S](t)$, which also explains the earlier relative success in applying linearization and graphical analysis to the initial velocity equation.

However, by using equation (3.A8) instead of (3.A6a) expands the range of reaction rates and provides a new kinetic equation, in the form of a logistic expression

$$-\frac{1}{V_{\max}} \frac{d}{dt}[S](t) = 1 - e^{-\frac{[S](t)}{K_M}} \tag{3.A10}$$

based on probability and derived from equations and (3.A1), (3.A5), and (3.A8).

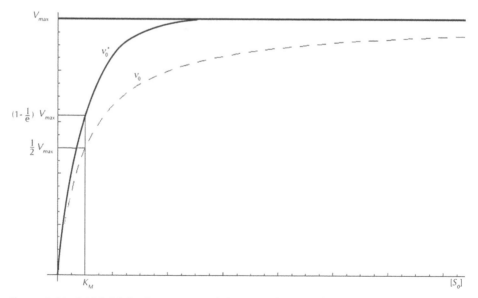

Figure 3.A1. Initial Michaelis–Menten and logistic velocities plotted against initial substrate concentration for the E-S mono-substrate enzymic reaction. The dashed curve corresponds to the Michaelis–Menten equation (3.A6) while the continuous thick curve represents its logistic generalization from (3.A10): $v_0^* = V_{\max}[1 - \exp(-[S_0]/K_M)]$.

At initial conditions, logistic equation (3.A10) gives an initial velocity of reaction (v_0^*) that is uniformly higher than that calculated by Michaelis–Menten (3.A6a) at all initial concentrations of the substrate, except for the case where $[S_0] \rightarrow 0$, when both are zero, see Figure 3.A1. To test whether the logistic kinetic equation (3.A10), which is a natural generalization of the Michaelis–Menten equation, may provide a workable analytical solution in an elementary form we first integrate it under the form

$$\int_{[S_0]}^{[S](t)} \frac{d[S](t)}{\exp(-[S](t)/K_M)-1} = \int_0^t V_{max}\, dt \qquad (3.A11)$$

generating the new equation to be solved:

$$[S_0]-[S](t)+K_M \ln\left(e^{-\frac{[S_0]}{K_M}}-1\right)-K_M \ln\left(e^{-\frac{[S](t)}{K_M}}-1\right) = V_{max}t. \qquad (3.A12)$$

This can be solved exactly by substituting

$$\phi([S](t)) = \frac{[S](t)}{K_M} \qquad (3.A13)$$

into (3.A12) to get the simple equation:

$$-\phi([S](t)) - \ln\left(e^{-\phi([S](t))}-1\right) = \psi(t) \qquad (3.A14)$$

where we have also introduced the functional notation:

$$\psi(t) = \frac{1}{K_M}(V_{max}t - [S_0]) - \ln\left(e^{-\frac{[S_0]}{K_M}}-1\right). \qquad (3.A15)$$

Now, the exact solution of equation (3.A14) takes the logistic expression:

$$\phi([S](t)) = \ln\left(1-e^{-\psi(t)}\right). \qquad (3.A16)$$

Finally, substituting function (3.A15) into expression (3.A16) gives the logistic progress curve for substrate consumption in an analytically elementary form:

$$[S]_L(t) = K_M \ln\left(1+e^{-\frac{V_{max}t}{K_M}}\left(e^{\frac{[S_0]}{K_M}}-1\right)\right). \qquad (3.A17)$$

This time-dependent solution (3.A17) substitutes an elementary logarithmic dependency for the W-Lambert function, viewed so far as the best compact form solution of the Michaelis–Menten monosubstrate enzymic kinetics [13–15, 72, 73]. It is nevertheless remarkable that the solution of a generalized logistic kinetic version of the Michaelis–Menten instantaneous equation provides an analytically exact solution.

KEYWORDS

- **Hydrophobicity**
- **Microorganisms**
- **Steric effects**

PERMISSIONS

This Chapter was previously published in Advances in Chemical Bonding Structures, Mihai V. Putz (Ed.) as Designing a Spectral Structure-Activity Ecotoxico-Logistical Battery (Lacrămă A.M., Putz M.V., Ostafe V.) by Transworld Research Network, Kerala, India (2008), ISBN: 978-81-7895-306-9, pp. 389-419.

Chapter 4

ESIP (Element Specific Influence Parameter) SPECTRAL-SAR Molecular Activity Combined Models Toward Inter-species Toxicity Assessment

Sergiu Andrei Chicu and Mihai V. Putz

INTRODUCTION

Aiming to provide a unified picture of *computed* activity—quantitative structure-activity relationships, the so called Köln (*ESIP*—Element *S*pecific Influence Parameter) model for activity and Timişoara (SPECTRAL-SAR) formulation of QSAR were pooled in order to assess the toxicity modeling and inter-toxicity correlation maps for aquatic organisms against paradigmatic organic compounds. The Köln *ESIP* model for estimation of a compound toxicity is based on the experimental measurement expressing the direct action of chemicals on the organism *Hydractinia echinata* so that the structural influence parameters are reflected by the metamorphosis degree itself. As such, the calculation of the structural parameters is absolutely necessary for correct evaluation and interpretation of the evolution of M(easured) and the C(computed) values. On the other hand, the Timişoara SPECTRAL-SAR analysis offers correlation models and paths for *Hydractinia echinata* (*H.e.*) species, as well as for four other different organisms with which the toxicity may be inter-changed by means of the same mechanism of action induced by certain common chemicals.

Directly and without delay inclusion of chemically artifacts in the biological cycle are due in the first line to solubility; from these, all less soluble, that is, those set down as sediments, suffer with the time various transformations with formation of new derivatives and with other possibilities of implication in the same natural biological cycle [1]. However, in the all of the cases the principal area of the accumulation is mainly the shallow marine water where the effects can be detected immediately to intimation or pursued in the time with different investigation methods.

Hydractinia echinata, as an organism living in the European and North-American coastal waters, could be directly affected by the presence of chemical derivatives through interruption of the evolution cycle at the level of the larva to polyp metamorphosis [2].

The testing of many anticonvulsants through which it was established that the order of influence is identical to that obtained through treatment of the embryo *in vitro* [3] was first achieved by use of the *Hydractinia echinata* metamorphosis stage for monitoring toxicity problems. The research continued by establishing various relation-

---

Chicu S.A., Putz M.V. "Köln-Timişoara Molecular Activity Combined Models toward Interspecies Toxicity Assessment", *International Journal of Molecular Sciences*, 10(10) (2009) 4474-4497.

ships between structure and reactivity of oil and oil products, alkanes, cycloalkanes, aromatic compounds [4], as well as for a series of hydrocarbon derivatives, aliphatic alcohols, aliphatic amines, aminoalcohols [5], or phenols [6].

The *Hydractinia echinata* test-system was already demonstrated to be applicable for very different series of derivatives or products, including pharmaceutical products for dentistry, natural extracts, detergents, dyes, and so forth. Their interactions on living cells from measured (M) values for simple organic molecules by means of the introduced *ESIP*-parameters models the molecular substructures for their computed (C) toxicity of containing substances represent the essence of the so called "Köln model" [5]. On the other side, the recently developed so called *SPECTRAL-SAR* as the "Timişoara QSAR model" allows for mechanistic description of the molecular specific actions throughout combined reactivity-activity paths of interactions [7–10].

In this context, the present endeavor combines ESIP and SPECTRAL-SAR models for advancing a sort of "absolute" analysis of ecotoxicity employing the computed activities of their spectral correlation, respectively, for an inter-species analysis for a common set of compounds. As such, having at hand a complex method providing both the organisms' toxicity activity (*ESIP*), without the need to undertake extensive experiments for measuring them, as well as the mechanistically revealed path of molecular action (SPECTRAL-SAR) may constitute an advancement in ecotoxicological assessments through computational design and reasoning. This way, the *in silico* methods will eventually reveal the mechanisms of toxicity for a given set of toxicants and environmental hazards, while lowering the experimental costs.

BACKGROUND MODELS

Kuln ESIP Model for Biological Activity

We determinate *ESIP*-parameters based on the measured values basis $Mlog(1/MRC_{50})$ [Mol/L] in order to calculate toxicity values $Clog(1/MRC_{50})$ [Mol/L] for untested derivatives [5]. The molecular structures have always saturated hydrocarbon or aromatic substructures, so the first *ESIP*-parameter corresponds to saturated-carbon *ESIP*c-sat, followed by the aromatic-carbon *ESIP*c-ar, and *ESIP*-organic function (alcohol, amino, etc.). In the case of saturated hydrocarbons the *ESIP*c-sat have an average value of 0.50 log units, calculated on the basis of measured values M and saturated carbon numbers C.

In this way, the toxicity of not tested compounds can be calculated with the following assumptions:

  i. The toxicity of a compound can be subdivided into that of components (*ESIP*'s) in such a way that the sum of these components results in the total toxicity value;

  ii. These components (*ESIP*'s) are identical in different substances;

  iii. The *ESIP*'s components have a dynamical value (they depend on the determined number or are derived from newly available data) for one organism and a test-system, while varying for different test-systems. However, if a deviation between the measured M and the calculated C values is observed, there is an indication of an overlooked interaction between different parts of the molecule, or may indicate an activity of a substance specific for a certain bio-chemical pathway.

Note that somewhat similar studies were examined at the inter-species toxicity level by the aid of data bases centered on a given species [11], although this limits the possibility to dynamically extend the molecular group toxicity from one organism to other [12, 13], as the *ESIP* method is able to do.

Timişoara SPECTRAL-SAR Model

Since QSAR models aim at correlations between concerned (congener) molecular structures and measured (or otherwise evaluated) activities, it appears naturally that the *structure* part of the problem be accommodated within the quantum theory and of its formalisms. In fact, there are few quantum characters that we are using within the present approach:

- Any molecular structural state (dynamical, since undergoes interactions with organisms) may be represented by a $|ket\rangle$ *state vector*, in the abstract Hilbert space, following the $\langle bra|ket\rangle$ Dirac formalism [14]; such states are to be represented by any reliable molecular index, or, in particular in our study by hydrophobicity $|LogP\rangle$, polarizability $|POL\rangle$, and total optimized energy $|E_{tot}\rangle$, just to be restrained only the so called Hansch parameters, usually employed for accounting the diffusion, electrostatic, and steric effects for molecules acting on organisms' cells, respectively.

- The (quantum) *superposition principle* assuring that the various linear combinations of molecular states map onto the resulting state, here interpreted as the bio-, eco-, or toxico-logical activity, that is, $|Y\rangle = |Y_0\rangle + C_{LogP}|LogP\rangle + C_{POL}|POL\rangle + ...$, with $|Y_0\rangle$ meaning the free or unperturbed activity (when all other influences are absent).

- The *orthogonalization feature* of quantum states, a crucial condition providing that the superimposed molecular states generates *new* molecular state (here quantified as the organism activity); analytically, the orthogonalization condition is represented by the $\langle bra|ket\rangle$ scalar product of two envisaged states (molecular indices); if it is evaluated to zero value, that is, $\langle bra|ket\rangle = 0$, then the convoluted states are said to be orthogonal (zero-overlapping) and the associate molecular descriptors are considered as independent, therefore suitable to be assumed as eigen-states (of a *spectral* decomposition) in the resulted activity state, while quantified by the degree their molecular indices enter the activity correlation. Further details on scalar product and related properties are given in Appendix A1, whereas in what follows the spectral-based SAR correlation method (thereby called as SPECTRAL-SAR) is resumed.

Note that since molecular states are usually represented by *ket* vectors which are a generalization of custom (classical) vectors, all formalisms are consistently developed accordingly. In this regard, the *bra-ket* formalism is more than a simple notation—it is indeed a reliable formalism since, for instance, it differentiates between the dual and direct spaces the *bra-* and *ket-* vectors are attributed to, respectively, with insightful consequences for the space-time evolution of a system—a matter not conveyed by classical simple vectorial notation. However, it is not a complication of reality but a close representation of it: the molecular descriptors belong to a given molecular state that *has* to be included as a component of the quantum (*ket*) vectors carrying the specific structural information—a feature not fulfilled by simple classical vectors. Therefore, the adopted vectorial formalism goes beyond the simple notation—each time when we write a *ket* vector represented by a structural index we see in fact a generalized electronic (for a hyper-molecular) state, defined as the global state collecting one descriptor' values for all concerned congener molecules.

Now, a set of N molecules studied against observed/recorded/measured biological activity is represented by means of their M—structural indicators (the states); all the $N \times M$ input information may be expressed by the vectors-columns of the Table 4.1 and correlated upon the generic scheme of equations (4.1a)–(4.1d):

$$\left| Y_{OBS(ERVED)} \right\rangle = \left| Y_{PRED(ICTED)} \right\rangle + \left| prediction \ \ error \right\rangle \tag{4.1a}$$

$$= b_0 \left| X_0 \right\rangle + b_1 \left| X_1 \right\rangle + ... + b_k \left| X_k \right\rangle + ... + b_M \left| X_M \right\rangle + \left| prediction \ \ error \right\rangle, \tag{4.1b}$$

where the vector $\left| X_0 \right\rangle = \left| 1 \ \ 1 \ \ ... \ \ 1_N \right\rangle$ was added to account for the free activity term.

Table 4.1. The vectorial (molecular) descriptors in a SPECTRAL-SAR analysis represented as states, within the Hilbert N-dimensional space of investigated molecules.

Activity	Structural predictor variables					
$\left\| Y_{OBS(ERVED)} \right\rangle$	$\left\| X_0 \right\rangle$	$\left\| X_1 \right\rangle$	...	$\left\| X_k \right\rangle$	...	$\left\| X_M \right\rangle$
$y_{1\text{-}OBS}$	1	x_{11}	...	x_{1k}	...	x_{1M}
$y_{2\text{-}OBS}$	1	x_{21}	...	x_{2k}	...	x_{2M}
...	...	...	...	...	...	...
$y_{N\text{-}OBS}$	1	x_{N1}	...	x_{Nk}	...	x_{NM}

In order for equation (4.1b) to represent a reliable model of the given activities, the hyper-molecular states (indices) assumed should constitute an orthogonal set, having this constraint a consistent quantum mechanical basis, as above described. However, unlike other important studies addressing this problem [15–17], the present SPECTRAL-SAR [7] assumes the prediction error vector as being orthogonal to all others:

$$\left\langle Y_{PRED} \right| prediction \ \ error \right\rangle = 0 \tag{4.1c}$$

since it is not known *a priori* any correlation is made. Moreover, equations (4.1a), (4.1b), and (4.1c) imply that the prediction error vector has to be orthogonal on all known descriptors (states) of predicted activity:

$$\left\langle X_{i=\overline{0,M}} \right| prediction \ \ error \right\rangle = 0, \tag{4.1d}$$

assuring therefore the reliability of the present $\left| ket \right\rangle$ states approach. In other terms, conditions (4.1c) and (4.1d) agree with equation (4.1a) in the sense that the prediction vector and the prediction activity $\left|Y_{PRED}\right\rangle$ (with all its sub-intended states $\left|X_{\overline{i=0,M}}\right\rangle$) belong to disjoint (thus orthogonal) Hilbert (sub)spaces; or, even more, one can say that the Hilbert space of the observed activity $\left|Y_{OBS}\right\rangle$ may be decomposed into a predicted and error independent Hilbert sub-spaces of states.

Therefore, within Timişoara SPECTRAL-SAR procedure the very first step consists in orthogonalization of *prediction error* on the predicted activity and on its predictor states, while the remaining algorithm does not seek to optimize the minimization of errors, but for producing the ideal correlation between $\left|Y_{PRED}\right\rangle$ and the given descriptors $\left|X_{\overline{i=0,M}}\right\rangle$.

Next, the Gram-Schmidt orthogonalization scheme is applied through construction of the appropriate set of descriptors by means of the consecrated iteration [16, 18, 19]:

$$\left|\Omega_0\right\rangle = \left|X_0\right\rangle, \tag{4.2a}$$

$$\left|\Omega_k\right\rangle = \left|X_k\right\rangle - \sum_{i=0}^{k-1} r_i^k \left|\Omega_i\right\rangle, \tag{4.2b}$$

$$r_i^k = \frac{\left\langle X_k\middle|\Omega_i\right\rangle}{\left\langle \Omega_i\middle|\Omega_i\right\rangle}, k = \overline{1,M}, \tag{4.2c}$$

providing the orthogonal correlation:

$$\left|Y_{PRED}\right\rangle = \omega_0\left|\Omega_0\right\rangle + \omega_1\left|\Omega_1\right\rangle + ... + \omega_k\left|\Omega_k\right\rangle + ... + \omega_M\left|\Omega_M\right\rangle, \tag{4.3a}$$

$$\omega_k = \frac{\left\langle \Omega_k\middle|Y_{PRED}\right\rangle}{\left\langle \Omega_k\middle|\Omega_k\right\rangle}, k = \overline{0,M}. \tag{4.3b}$$

Remarkably, while available studies dedicated to the orthogonality problem usually stop at this stage, the SPECTRAL-SAR uses it to provide the solution for the original sought correlation of equation (4.1b)—having the prediction error vector orthogonal to the predicted activity and all its predictor states of Table 4.1. This can be wisely achieved through grouping equations (4.2) and (4.3) so that the system of all descriptors of Table 4.1 is now written in terms of orthogonal descriptors:

$$\left\{\begin{array}{l} \left|Y_{PRED}\right\rangle = \omega_0\left|\Omega_0\right\rangle + \omega_1\left|\Omega_1\right\rangle + ... + \omega_k\left|\Omega_k\right\rangle + ... + \omega_M\left|\Omega_M\right\rangle \\[2mm] \left|X_0\right\rangle = 1\cdot\left|\Omega_0\right\rangle + 0\cdot\left|\Omega_1\right\rangle + ... + 0\cdot\left|\Omega_k\right\rangle + ... + 0\cdot\left|\Omega_M\right\rangle \\[2mm] \left|X_1\right\rangle = r_0^1\left|\Omega_0\right\rangle + 1\cdot\left|\Omega_1\right\rangle + ... + 0\cdot\left|\Omega_k\right\rangle + ... + 0\cdot\left|\Omega_M\right\rangle \\[2mm] .... \\[2mm] \left|X_k\right\rangle = r_0^k\left|\Omega_0\right\rangle + r_1^k\left|\Omega_1\right\rangle + ... + 1\cdot\left|\Omega_k\right\rangle + ... + 0\cdot\left|\Omega_M\right\rangle \\[2mm] .... \\[2mm] \left|X_M\right\rangle = r_0^M\left|\Omega_0\right\rangle + r_1^M\left|\Omega_1\right\rangle + ... + r_k^M\left|\Omega_k\right\rangle + ... + 1\cdot\left|\Omega_M\right\rangle \end{array}\right. \tag{4.4}$$

According with a well known algebraic theorem, the system (4.4) has no trivial solution if and only if the associated extended determinant vanishes; this way the SPECTRAL-SAR determinant features the form [7]:

$$
\begin{vmatrix}
\|Y_{PRED}\rangle & \omega_0 & \omega_1 & \dots & \omega_k & \dots & \omega_M \\
|X_0\rangle & 1 & 0 & \dots & 0 & \dots & 0 \\
|X_1\rangle & r_0^1 & 1 & \dots & 0 & \dots & 0 \\
\dots & \dots & \dots & \dots & \dots & \dots & \dots \\
|X_k\rangle & r_0^k & r_1^k & \dots & 1 & \dots & 0 \\
\dots & \dots & \dots & \dots & \dots & \dots & \dots \\
\|X_M\rangle & r_0^M & r_1^M & \dots & r_k^M & \dots & 1
\end{vmatrix} = 0 . \qquad (4.5)
$$

Now, when the determinant of equation (4.5) is expanded on its first column, and the result is rearranged so that to have $|Y_{PRED}\rangle$ on left side and the rest of states/indicators on the right side the sought QSAR solution for the initial observed-predicted correlation problem of equation (4.1a) is obtained under the SPECTRAL-SAR vectorial expansion (from where the "spectral" name is justified) without the need to minimize the predicted error vector anymore, being this stage absorbed in its orthogonal behavior with respect to the predicted activity.

In fact, the SPECTRAL-SAR procedure uses the double conversion idea: one forward, from the given problem of equations (4.1a)–(4.1d) to the orthogonal one of equation (4.3) in which the error vector has no manifestation; and a backwards one, from the orthogonal to the real descriptors by employing the system (4.4) determinant (4.5) expansion as the QSAR solution.

It is worth stressing that the present QSAR/SPECTRAL-SAR equations are totally delivered from the (analytical) determinant (4.5) and not computationally restricted to the inverse matrix product as prescribed by the fashioned statistical Pearson approach [20]. Moreover, the SPECTRAL-SAR algorithm is invariant also upon the order of descriptors chosen in orthogonalization procedure, providing equivalent determinants no matter how its lines are re-derived, an improvement that was not previously achieved by other available orthogonalization techniques [15, 17].

However, besides the effectiveness of the SPECTRAL-SAR methodology in reproducing the old-fashioned multi-linear QSAR analysis [7, 21], one of its advantages concerns on the possibility of introducing the so called (*vectorial*) *norms* (see Appendix A1) associated with either *experimental* (measured or observed) or *predicted* (computed) activities:

$$
\left\| Y_{OBS\,/\,PRED}\rangle \right\| = \sqrt{\langle Y_{OBS\,/\,PRED} | Y_{OBS\,/\,PRED}\rangle} = \sqrt{\sum_{i=1}^{N} y_{i-OBS\,/\,PRED}^2} ; \qquad (4.6)
$$

They provide a unique assignment of a number to a specific type of correlation, that is, by performing a sort of final quantification of the models. Nevertheless, the activity norm given in equation (4.6) opens the possibility of replacing the classical statistical correlation factor [21]:

$$R \equiv r_{STATISTIC} = \sqrt{1 - \frac{\sum\limits_{i=1}^{N}\left(y_{i-OBS} - y_{i-PRED}\right)^2}{\sum\limits_{i=1}^{N}\left(y_{i-OBS} - \frac{1}{N}\sum\limits_{i=1}^{N}y_{i-OBS}\right)^2}},\qquad (4.7)$$

with a new index of correlation, introduced as the so called *algebraic SPECTRAL-SAR correlation factor* (or *R*-algebraic, shorthanded as *RA*) through the ratio of the predicted to observed norms [22, 23]:

$$RA \equiv r_{ALGEBRAIC} = \sqrt{\frac{\sum\limits_{i=1}^{N}y_{i-PRED}^2}{\sum\limits_{i=1}^{N}y_{i-OBS}^2}} = \frac{\left\| Y_{PRED}\right\rangle \right\|}{\left\| Y_{OBS}\right\rangle \right\|};\qquad (4.8)$$

It has the meaning of realization probability with which a certain predicted model approaches the observed activity throughout all of the employed molecules (in the hyper-molecular states of activities), see Appendix (4.6.A2).

With this interpretation the algebraic correlation conceptually departs from the statistical one in that the later accounts on the degree with which each computed individual molecular activity approaches the *mean* activity of the N-molecules, while the first evaluates the (hyper-molecule) degree of overlap of predicted to observed activities' norms (viewed as the "amplitudes" of molecular-organism interaction's intensity). In this respect there seems that the algebraic analysis is more suited to environmental studies in which the *global* rather than *local* effect of a series of toxicants is evaluated on specific species and organisms.

In fact, this new correlation factor definition compares the vectorial lengths of the predicted activity against the measured one, thus being an indicator of the extent with which certain computed property or activity approaches the "length" of the observed quantity.

However, it was already shown that the algebraic correlation factor of equation (4.8) furnishes higher and more insightful values than its statistical counterpart in a systematical manner [21, 24], thus advancing it as the ideal tool for correlation analysis on a shrink interval of data analysis where the statistical meaning is naturally lost.

Even more, in the terms of the "quantum spectral" formalism, one can say that algebraic investigation provides the "excited" states of an activity modeling, while the statistical approach deals with "ground state" or lower states of correlation. Consequently, for completeness, a proper quest of structure-activity models should include both of these stages of molecular SAR modeling.

Going further toward extracting the mechanistic information from the SPEC-TRAL-SAR norms and correlation factors we can further advance the so called *least path principle*:

$$\delta[A_1, A_2, ..., A_M] = \delta\left(\sum_{i=1}^{M-1}[A_i, A_{i+1}]\right) = 0 , \tag{4.9}$$

applied upon successively connected models with different correlation dimensions: it starts from 1-dimension with a single structural indicator correlation, say A_1, until the models with maximum factors of correlation, say A_M —that is, containing M number of indicators, see Table 4.1) [7–10]. Since each of these models is now characterized by its predicted activity norm $\left\|Y_{PRED}\right\rangle\right\|$ along the algebraic (RA) and/or statistical (R) correlation factors, the elementary paths of equation (4.9) are constructed as the Euclidian measure between two consecutive models (endpoints) [7–10, 22–24]:

$$[A_1, A_2] = \sqrt{\left(\left\|Y_{PRED}^{A_2}\right\rangle\right\| - \left\|Y_{PRED}^{A_1}\right\rangle\right\|\right)^2 + \left(r_{\substack{ALGEBRAIC \\ STATISTIC}}^{A_2} - r_{\substack{ALGEBRAIC \\ STATISTIC}}^{A_1}\right)^2} . \tag{4.10}$$

It is noteworthy that the formal equation (4.9) has to be read as searching for paths' combination on the left side providing minimum value in the right side; it is practiced as the tool for deciding the hierarchy along all (ergodic) possible end-point linked paths with the important consequence of picturing the mechanistic and causal evolution of structural influences that trigger the observed effects.

This methodology was successfully applied in ecotoxicology [7, 8, 24] and for designing the behavior of the species interactions within a test battery [23], promising to furnish adequate framework also for the present (and future) inter-species analysis.

SPECTRAL-SAR RESULTS

Data of Table 4.2 are modeled as QSARs for each species in both Mlog and Clog modes, with the help of SPECTRAL-SAR determinant (4.5), while reporting the algebraic norms and correlation computed upon equations (4.6) and (4.8), respectively, side-by-side with the statistical correlation coefficients of equation (4.7). The results are listed in Tables 4.3 and 4.4 for employed Mlog and Clog-*ESIP* data of Table 4.2, respectively. However, in order to assure the reliability for the computed models the so called Topliss–Costello rule was considered, that is, building models with about five times ratio of activity points with respect to the number of correlating/structural variables [25].

Aiming to provide the mechanistic maps of actions for the targeted species, the minimization principle of spectral paths given by equations (4.9) and (4.10) is considered among all possible ways of connecting endpoints from each category of models (i.e., with one, two, or three dependency factors). The Tables 4.5 and 4.6 present all these endpoints' paths for Mlog and Clog activities, computed upon equations (4.6)–(4.8) and (4.10) through processing the data of Tables 4.3 and 4.4, respectively.

Table 4.2. The measured Mlog(1/MRC50) and ESIP-computed Clog(1/MRC50) toxicities for *Hydractinia echinata* and other organisms: for compounds nos. 2–7 from Ref. [5], for compounds nos. 13–21 from Ref. [6], new data for the rest; the Hansch molecular parameters as hydrophobicity (LogP), polarizability (POL), and the steric optimized total energy (E_{tot}) were computed by HyperChem environment [26].

No.	Compound	Species Toxicities |Y>										Structural Parameters		
		Hydractinia echinata		*Tetrahymena pyriformis*		*Pimephales promelas*		*Vibrio fischeri*		*Daphnia magna*		|X_1> =Log P	|X_2> =POL (4.A3)	|X_3> =E_{tot} (kcal/mol)
		Mlog	Clog	Mlog	Clog	Mlog	Clog	Mlog	Clog	Mlog	Clog			
1	Water	-1.23	-0.91									-0.51	1.41	-8038.2
2	Methanol	-0.22	-0.41	0.33	0.15	0.04	0.24					-0.27	3.25	-11622.9
3	Ethanol	0.02	0.09	0.59	0.60	0.51	0.74	0.11	1.11	0.93	0.25	0.08	5.08	-15215.4
4	1-Butanol	0.99	1.08	1.48	1.50	1.63	1.73	1.34	2.18	1.57	1.68	0.94	8.75	-22402.8
5	1,2,3-Propanetriol	0.34	0.37									-1.08	8.19	-33600.
6	Triphenylmethanol	5.69	5.27									4.87	32.23	-68532.5
7	1,10-Diaminodecane	3.26	2.91									1.48	21.83	-46754.2
8	2-Benzylpyridine	3.75	3.46	3.41	4.85							3.53	21.22	-43675.3
9	4-Benzylpyridine	4.08	3.46	3.68	4.85							3.75	21.22	-43676.8
10	4-Phenylpyridin	4.13	3.46	3.66	3.46	3.98	3.81	4.91	4.84			3.35	19.38	-40083.1
11	4-Toluidine	2.85	2.02	2.98	2.81	3.43	3.26					1.73	13.62	-28300.3
12	1,2-Dichlorobenzene	3.04	3.45	4.00	3.66	4.19	4.17					3.08	14.29	-36217.2
13	Phenol(3,15/2,66/2,85)*	2.89*	2.87	2.79	2.58	3.41	3.21	3.42	3.68	3.32	3.32	1.76	11.07	-27003.1
14	2-Methylphenol(3,18/3,24)*	3.21*	2.82	2.72	2.58	3.77	3.21	3.75	3.68	3.64	3.32	2.23	12.91	-30596.6
15	2,4,6-Trimethylphenol(3,19/4,00)*	3.60*	3.82	3.42	3.48	4.02	4.21	4.08	4.75	4.49	4.75	3.16	16.58	-37783.7
16	1,2-Dihydroxibenzene	5.11	5.11	3.75	3.47	4.08	4.08	3.54	3.54	4.68	4.24	1.48	11.71	-34396.4
17	2-Methoxyphenol(2,89/2,77)*	2.83*	3.28	2.49	2.54				3.29			1.51	13.54	-37974.4
18	1,4-Dihydroxybenzene(6,14/6,06)*	6.10*	6.10	3.47	3.59	6.40	6.40	6.42	6.42			1.48	11.71	-34395.8
19	t-Butylhydroquinone(5,05/5,00)*	5.30*	7.60		4.94		7.78	8.03	8.03			3.11	19.05	-48758.1

Table 4.2. *(Continued)*

No.	Compound	Species Toxicities |Y>										Structural Parameters		
		Hydractinia echinata		*Tetrahymena pyriformis*		*Pimephales promelas*		*Vibrio fischeri*		*Daphnia magna*		$\|X_1>$ = Log P	$\|X_2>$ = P O L (4.A3)	$\|X_3>$ =E$_{tot}$ (kcal/mol)
		Mlog	Clog	Mlog	Clog	Mlog	Clog	Mlog	Clog	Mlog	Clog			
20	1,2,3-Trihydroxibenzene	5.15	5.15	3.85	3.65							1.19	12.35	–41789.9
21	4(3',5'-dimethyl--3'-heptyl) phenol	7.65	6.81									4.87	25.75	–55742.
22	4-Chlorophenol	3.25	3.04	3.55	3.56	4.18	3.66	4.19	3.88	4.13	3.95	2.28	13	–35307.6
23	2,6-Diisopropylphenol	3.73	5.31		4.82		5.21		6.36		6.90	4.15	22.08	–48554.7
24	2-Aminophenol	3.15	3.04	3.94	2.93							0.98	12.42	–32098.6
25	2,4,6-Trinitrophenol	2.92	2.99	2.84				2.63	3.77			–4.17	16.59	–84472.1
26	Chloranil	5.15	–									1.12	18.51	–66928.2
27	Chloranilic acid	3.40	2.99									–0.48	15.93	–65113.6
28	4-Methoxyazobenzene	5.20	3.70									4.10	24.63	–59069.5

The M values represents the experimental results accomplished by different time interval with different generations of *H.e.* These results clearly prove the reproducibility of the test-system.

Table 4.3. Mlog-Spectral SPECTRAL-SAR results employing the molecular parameters and the *Hydractinia echinata* (*H.e.*), *Tetrahymena pyriformis* (*T.p.*), *Pimephales promelas* (*P.p.*), *Vibrio fisheri* (*V.f.*), and *Daphnia magna* (*D.m.*) toxicities of Table 4.2; the models are characterized either by algebraic norms and correlation factors (*RA*), computed upon the equations (4.6) and (4.8), and by Pearson statistical correlation (*R*) of equation (4.7), for all possible mono-, bi-, and all- end-points, respectively. The referential algebraic norms of the considered species were estimated with the aid of equation (4.6) from the Mlog input toxicity data of Table 4.2 as: $\||Y_{H.e.}>\|| = 20.8547$, $\||Y_{T.p.}>\|| = 13.2774$, $\||Y_{P.p.}>\|| = 12.8515$, $\||Y_{V.f.}>\|| = 12.1055$, $\||Y_{D.m.}>\|| = 9.31242$.

Mlog Model	Species	SPECTRAL-SAR Activity Equation	SPECTRAL-SAR Norm	RA (Algebraic)	R (Statistic)
|1>	H.e.	$\|Y_{H.e.}{}^{\|1>}> = 2.348 + 0.595 \|LogP>$	19.0572	0.9138	0.5912
	T.p.	$\|Y_{T.p.}{}^{\|1>}> = 2.526 + 0.267 \|LogP>$	12.642	0.9521	0.4446
	P.p.	$Y_{P.p.}{}^{\|1>}> = 1.402 + 1.071 \|LogP>$	12.1481	0.9453	0.6972
	V.f.	$Y_{V.f.}{}^{\|1>}> = 2.981 + 0.364 \|LogP>$	11.1235	0.9189	0.4396
	D.m.	$Y_{D.m.}{}^{\|1>}> = 1.192 + 1.208 \|LogP>$	9.09749	0.9769	0.8300
|2>	H.e.	$\|Y_{H.e.}{}^{\|2>}> = 0.022 + 0.221 \|POL>$	19.7048	0.9449	0.7597
	T.p.	$\|Y_{T.p.}{}^{\|2>}> = 0.72 + 0.168 \|POL>$	12.9074	0.9721	0.7267
	P.p.	$\|Y_{P.p.}{}^{\|2>}> = -0.109 + 0.29 \|POL>$	12.2254	0.9513	0.7357
	V.f.	$\|Y_{V.f.}{}^{\|2>}> = 0.121 + 0.262 \|POL>$	11.3472	0.9374	0.6092
	D.m.	$\|Y_{D.m.}{}^{\|2>}> = -0.759 + 0.355 \|POL>$	9.16099	0.9837	0.8832
|3>	H.e.	$\|Y_{H.e.}{}^{\|3>}> = 0.433 - 0.00007 \|E_{tot}>$	19.2139	0.9213	0.6355
	T.p.	$\|Y_{T.p.}{}^{\|3>}> = 1.669 - 3.6 \cdot 10^{-5} \|E_{tot}>$	12.6819	0.9551	0.4969
	P.p.	$\|Y_{P.p.}{}^{\|3>}> = -1.767 - 1.7 \cdot 10^{-4} \|E_{tot}>$	12.5439	0.9761	0.8785
	V.f.	$\|Y_{V.f.}{}^{\|3>}> = 2.755 - 1.89 \cdot 10^{-5} \|E_{tot}>$	10.926	0.9026	0.1982
	D.m.	$\|Y_{D.m.}{}^{\|3>}> = -1.826 - 1.75 \cdot 10^{-4} \|E_{tot}>$	9.26686	0.9951	0.9662
|1,2>	H.e.	$\|Y_{H.e.}{}^{\|1,2>}> = 0.206 + 0.163\|LogP> + 0.19 \|POL>$	19.7462	0.9468	0.7694
	T.p.	$\|Y_{T.p.}{}^{\|1,2>}> = 0.784 + 0.093\|LogP> + 0.152 \|POL>$	12.9228	0.9733	0.7398
	P.p.	$\|Y_{P.p.}{}^{\|1,2>}> = -0.324 - 0.191\|LogP> + 0.337 \|POL>$	12.2271	0.9514	0.7365
	V.f.	$\|Y_{V.f.}{}^{\|1,2>}> = -0.018 + 0.307 \|LogP> + 0.242 \|POL>$	11.5146	0.9512	0.7116
|1,3>	H.e.	$\|Y_{H.e.}{}^{\|1,3>}> = -0.296 + 0.541\|LogP> - 0.00007 \|E_{tot}>$	20.0182	0.9599	0.8307
	T.p.	$\|Y_{T.p.}{}^{\|1,3>}> = 0.433 + 0.413\|LogP> - 5.27 \cdot 10^{-5} \|E_{tot}>$	13.018	0.9805	0.8171
	P.p.	$\|Y_{P.p.}{}^{\|1,3>}> = -3.541 - 1.061\|LogP> - 2.96 \cdot 10^{-4} \|E_{tot}>$	12.646	0.9840	0.9203
	V.f.	$\|Y_{V.f.}{}^{\|1,3>}> = -0.512 + 0.82 \|LogP> - 8.1 \cdot 10^{-5} \|E_{tot}>$	11.6329	0.9610	0.7767

Table 4.3. (Continued)

Mlog Model	Species	SPECTRAL-SAR Activity Equation	SPECTRAL-SAR Norm	RA (Algebraic)	R (Statistic)
$\|2,3\rangle$	H.e.	$\|Y_{H.e.}^{\|2,3\rangle}\rangle = -0.134 + 0.193 \|POL\rangle - 0.00001 \|E_{tot}\rangle$	19.7224	0.9457	0.7638
	T.p.	$\|Y_{T.p.}^{\|2,3\rangle}\rangle = 0.704 + 0.163 \|POL\rangle - 2.18 \cdot 10^{-6} \|E_{tot}\rangle$	12.9078	0.9722	0.7270
	P.p.	$\|Y_{P.p.}^{\|2,3\rangle}\rangle = -2.269 - 0.262\|POL\rangle - 2.94 \cdot 10^{-4} \|E_{tot}\rangle$	12.6208	0.9820	0.9101
	V.f.	$\|Y_{V.f.}^{\|2,3\rangle}\rangle = 0.082 + 0.36 \|POL\rangle + 3.35 \cdot 10^{-5} \|E_{tot}\rangle$	11.4347	0.9446	0.6645
$\{\|1,2,3\rangle\}$	H.e.	$\|Y_{H.e.}^{\{\|1,2,3\rangle\}}\rangle = -0.259 + 0.979\|LogP\rangle - 0.214\|POL\rangle - 0.00013\|E_{tot}\rangle$	20.1085	0.9642	0.8502
	T.p.	$\|Y_{T.p.}^{\{\|1,2,3\rangle\}}\rangle = 0.456 + 0.773\|LogP\rangle - 0.185\|POL\rangle - 0.00011\|E_{tot}\rangle$	13.0541	0.9832	0.8447

Table 4.4. The same type of SPECTRAL-SAR models as those of Table 4.3, here for Clog data of Table 4.2. The referential algebraic norms of the considered species were estimated with the equation (4.6) from the Clog input toxicity data of Table 4.2 as: $\|IY_{H.e.}\rangle\| = 20.1051$, $\|IY_{T.p.}\rangle\| = 14.8984$, $\|IY_{P.p.}\rangle\| = 15.5929$, $\|IY_{V.f.}\rangle\| = 16.6682$, $\|IY_{D.m.}\rangle\| = 11.3438$.

Clog Model	Species	SPECTRAL-SAR Activity Equation	SPECTRAL-SAR Norm	RA (Algebraic)	R (Statistic)
$\|1\rangle$	H.e.	$\|Y_{H.e.}^{\|1\rangle}\rangle = 2.242 + 0.583 \|LogP\rangle$	18.1498	0.9027	0.5744
	T.p.	$\|Y_{T.p.}^{\|1\rangle}\rangle = 1.248 + 0.919 \|LogP\rangle$	14.6075	0.9805	0.8572
	P.p.	$\|Y_{P.p.}^{\|1\rangle}\rangle = 1.436 + 1.107 \|LogP\rangle$	14.7182	0.9439	0.7011
	V.f.	$\|Y_{V.f.}^{\|1\rangle}\rangle = 3.598 + 0.41 \|LogP\rangle$	15.6785	0.9406	0.4605
	D.m.	$Y_{D.m.}^{\|1\rangle}\rangle = 0.57 + 1.483 \|LogP\rangle$	11.2079	0.9880	0.9432
$\|2\rangle$	H.e.	$\|Y_{H.e.}^{\|2\rangle}\rangle = 0.242 + 0.201 \|POL\rangle$	18.5655	0.9234	0.6831
	T.p.	$\|Y_{T.p.}^{\|2\rangle}\rangle = -0.118 + 0.237 \|POL\rangle$	14.7122	0.9875	0.9108
	P.p.	$\|Y_{P.p.}^{\|2\rangle}\rangle = -0.092 + 0.29 \|POL\rangle$	14.871	0.9537	0.7604
	V.f.	$\|Y_{V.f.}^{\|2\rangle}\rangle = 0.111 + 0.298 \|POL\rangle$	16.1385	0.9682	0.7565
	D.m.	$\|Y_{D.m.}^{\|2\rangle}\rangle = -1.241 + 0.379 \|POL\rangle$	11.2605	0.9927	0.9655
$\|3\rangle$	H.e.	$\|Y_{H.e.}^{\|3\rangle}\rangle = 0.518 - 0.00007 \|E_{tot}\rangle$	18.16	0.9033	0.5773
	T.p.	$\|Y_{T.p.}^{\|3\rangle}\rangle = -1.176 - 1.27 \cdot 10^{-4} \|E_{tot}\rangle$	14.8013	0.9935	0.9544
	P.p.	$\|Y_{P.p.}^{\|3\rangle}\rangle = -1.597 - 1.64 \cdot 10^{-4} \|E_{tot}\rangle$	15.2359	0.9771	0.8882
	V.f.	$\|Y_{V.f.}^{\|3\rangle}\rangle = 2.546 - 4.51 \cdot 10^{-5} \|E_{tot}\rangle$	15.6221	0.9372	0.4106
	D.m.	$\|Y_{D.m.}^{\|3\rangle}\rangle = -2.546 - 1.94 \cdot 10^{-4} \|E_{tot}\rangle$	11.3184	0.9978	0.9896

Table 4.4. (Continued)

Clog Model	Species	SPECTRAL-SAR Activity Equation	SPECTRAL-SAR Norm	RA (Algebraic)	R (Statistic)
$\lvert 1,2\rangle$	H.e.	$\lvert Y_{H.e.}^{\lvert 1,2\rangle}\rangle = 0.488 + 0.217\,\lvert LogP\rangle + 0.16\,\lvert POL\rangle$	18.6415	0.9272	0.7014
	T.p.	$\lvert Y_{T.p.}^{\lvert 1,2\rangle}\rangle = -0.208 - 0.081\lvert LogP\rangle + 0.255\,\lvert POL\rangle$	14.7128	0.9875	0.9112
	P.p.	$\lvert Y_{P.p.}^{\lvert 1,2\rangle}\rangle = -1.038 - 0.931\lvert LogP\rangle + 0.509\,\lvert POL\rangle$	14.908	0.9561	0.7742
	V.f.	$\lvert Y_{V.f.}^{\lvert 1,2\rangle} = 0.228 + 0.188\,\lvert LogP\rangle + 0.268\,\lvert POL\rangle$	16.187	0.9711	0.7816
$\lvert 1,3\rangle$	H.e.	$\lvert Y_{H.e.}^{\lvert 1,3\rangle}\rangle = -0.134 + 0.522\,\lvert LogP\rangle - 0.00006\,\lvert E_{tot}\rangle$	18.9449	0.9423	0.7708
	T.p.	$\lvert Y_{T.p.}^{\lvert 1,3\rangle}\rangle = -0.859 + 0.219\,\lvert LogP\rangle - 1.04\cdot 10^{-4}\,\lvert E_{tot}\rangle$	14.8152	0.9944	0.9611
	P.p.	$\lvert Y_{P.p.}^{\lvert 1,3\rangle}\rangle = -3.524 - 1.327\lvert LogP\rangle - 3.1\cdot 10^{-4}\,\lvert E_{tot}\rangle$	15.4088	0.9882	0.9437
	V.f.	$\lvert Y_{V.f.}^{\lvert 1,3\rangle}\rangle = -0.12 + 0.713\,\lvert LogP\rangle - 8.42\cdot 10^{-5}\,\lvert E_{tot}\rangle$	16.2777	0.9766	0.8267
$\lvert 2,3\rangle$	H.e.	$\lvert Y_{H.e.}^{\lvert 2,3\rangle}\rangle = 0.093 + 0.175\,\lvert POL\rangle - 0.00001\,\lvert E_{tot}\rangle$	18.5804	0.9242	0.6868
	T.p.	$\lvert Y_{T.p.}^{\lvert 2,3\rangle}\rangle = -1.045 + 0.062\,\lvert POL\rangle - 9.77\cdot 10^{-5}\lvert E_{tot}\rangle$	14.8118	0.9942	0.9594
	P.p.	$\lvert Y_{P.p.}^{\lvert 2,3\rangle}\rangle = -2.243 - 0.355\,\lvert POL\rangle - 3.28\cdot 10^{-4}\,\lvert E_{tot}\rangle$	15.3717	0.9858	0.9320
	V.f.	$\lvert Y_{V.f.}^{\lvert 2,3\rangle} = 0.2 + 0.337\,\lvert POL\rangle + 1.65\cdot 10^{-5}\,\lvert E_{tot}\rangle$	16.1548	0.9692	0.7650
$\{\lvert 1,2,3\rangle\}$	H.e.	$\lvert Y_{H.e.}^{\{\lvert 1,2,3\rangle\}}\rangle = -0.166 + 1.229\lvert LogP\rangle - 0.351\lvert POL\rangle - 0.00017\lvert E_{tot}\rangle$	19.1684	0.9534	0.8188
	T.p.	$\lvert Y_{T.p.}^{\{\lvert 1,2,3\rangle\}}\rangle = -0.871 + 0.199\lvert LogP\rangle + 0.008\lvert POL\rangle - 0.0001\lvert E_{tot}\rangle$	14.8153	0.9944	0.9611

Table 4.5. Synopsis of the statistic and algebraic values of paths connecting the SPECTRAL-SAR models for *Hydractinia echinata* (*H.e.*) and *Tetrahymena pyriformis* (*T.p.*) in the Mlog/Clog and algebraic/statistical computational frames of Tables 4.3 and 4.4. The primary, secondary, and tertiary—the so called *alpha* (α), *beta* (β), and *gamma* (γ)—paths are indicated according to the least path principle in spectral norm-correlation space, respectively.

Species	H.e.				T.p.			
Method	Mlog		CLog		Mlog		Clog	
Paths	Algebraic	Statistic	Algebraic	Statistic	Algebraic	Statistic	Algebraic	Statistic
$\lvert 1\rangle\!\rightarrow\!\lvert 1,2\rangle\!\rightarrow\!\lvert 1,2,3\rangle$	1.05246	*1.08283[γ]*	1.01981	*1.0476[γ]*	0.41325	*0.575439[γ]*	0.208278	0.232353
$\lvert 1\rangle\!\rightarrow\!\lvert 1,3\rangle\!\rightarrow\!\lvert 1,3,2\rangle$	**1.05246[γ]**	1.08273	**1.01981[γ]**	1.04754	**0.41325[γ]**	0.574673	**0.208278[γ]**	*0.232342[γ]*

Table 4.5. *(Continued)*

Species	H.e.				T.p.			
Method	Mlog		CLog		Mlog		Clog	
Paths	Algebraic	Statistic	Algebraic	Statistic	Algebraic	Statistic	Algebraic	Statistic
|1>→|2,3>→|1,2,3>	1.05246	1.08284	1.01981	1.0476	0.41325	0.575534	0.208278	0.232342
|2>→|1,2>→|2,1,3>	0.404191	0.413755	0.603637	0.61798	0.147067	0.188257	0.103313[β]	*0.114674[β]*
|2>→|1,3>→|2,1,3>	0.404191	0.413756	0.603637	0.617987	0.147067	0.188265	0.103313	0.114674
|2>→|2,3>→|2,3,1>	**0.404191[α]**	*0.413754[α]*	**0.603637[α]**	*0.617972[α]*	**0.147067[α]**	*0.188246[α]*	0.103313	0.114674
|3>→|1,2>→|3,1,2>	**0.89559[β]**	0.920041	**1.00961[β]**	1.03703	**0.373261[β]**	0.510175	0.191347	0.212443
|3>→|1,3>→|3,1,2>	0.89559	*0.919987[β]*	1.00961	*1.03697[β]*	0.373261	*0.509664[β]*	0.0140336	0.0155182
|3>→|2,3>→|3,2,1>	0.89559	0.920044	1.00961	1.03703	0.373261	0.510232	**0.0140336[α]**	**0.0155182[α]**

Table 4.6. The same type of information and analysis as in Table 4.5, here for *Pimephales promelas* (*P.p.*) and *Vibrio fisheri* (*V.f.*) species.

Species	P.p.				V.f.			
Method	Mlog		CLog		Mlog		Clog	
Paths	Algebraic	Statistic	Algebraic	Statistic	Algebraic	Statistic	Algebraic	Statistic
|1>→|1,2>	0.0792073	0.0881801	0.190201	0.2034	**0.392451[β]**	*0.476398[β]*	**0.509418[β]**	*0.601392[β]*
|1>→|1,3>	**0.499343[γ]**	*0.545515[γ]*	**0.692013[γ]**	*0.731962[γ]*	0.511148	0.61083	0.600329	0.702271
|1>→|2,3>	0.474093	0.518389	0.654817	0.69307	0.312213	0.383883	0.477152	0.565317
|2>→|1,2>	**0.00166893[α]**	*0.0018496[α]*	**0.0371086[α]**	*0.0395137[α]*	0.167993	0.196303	0.0485269	0.0545366
|2>→|1,3>	0.421805	0.459267	0.538921	0.568184	0.28669	0.331225	0.139438	0.155864
|2>→|2,3>	0.396554	0.432134	0.501725	0.529282	**0.0877552[α]**	*0.103489[α]*	**0.0162612[α]**	*0.0183134[α]*
|3>→|1,2>	0.3177	0.347114	0.328541	0.347126	0.590616	0.781086	0.565916	0.675813
|3>→|1,3>	0.102435	0.11035	0.173271	0.181596	**0.709313[γ]**	*0.913454[γ]*	**0.656827[γ]**	*0.776511[γ]*
|3>→|2,3>	**0.0771849[β]**	*0.0832042[β]*	**0.136075[β]**	*0.142683[β]*	0.510379	0.690032	0.53365	0.639803

However, in order to identify the shortest paths in each category of endpoint connections, according with prescription given by equation (4.9), the following rules are applied:

i.   The first choice is the overall minimum path, in a certain column of Tables 4.5 and 4.6 (either for statistical or algebraically correlation);

ii. If the overall minimum is reached by many equivalent paths (as is the case of Mlog-algebraic column for *H.e.* in Table 4.4.5, for instance) the minimum path will be considered that one connecting the starting endpoint with the closest endpoint in the sense of norms (as is for *H.e.*/Mlog the norm of |2> state the closest to the norm of |2,3> state, as compared with |1,2> and |1,3>, see, e.g., SPECTRAL-SAR norm column of Table 4.3);

iii. The overall minimum path will set the dominant hierarchical path in assessing the mechanistically mode of action toward the given/measured activity; it is called as *the alpha path* (α);

iv. Once the alpha path has been set the next minimum path will be looked for in such a way that the new starting endpoint is different from that one already involved in the alpha path (i.e., if in the established *alpha* path for *H.e.*/Mlog the starting model correspond to the |2> state, the next path to be identified will originate either on models/states |1> or |3>);

v. The remaining minimum paths are identified on the same rules as before and will be called like *beta* and *gamma paths*, β and γ, respectively;

vi. At the end of this procedure each mode of action is to be "touched" only one, excepting the final endpoint state {|1,2,3>} that can present *degeneracy*, that is., may be found with the same influence at the end of various paths, herein called as *degenerate paths* (e.g., the states |1,2,3>, |2,1,3>, and |3,1,2> in the case of *Hydractinia echinata* and *Tetrahymena pyriformis* at their ending toxicity paths of Table 4.5); Yet, such behavior may leave with the important idea the *degenerate paths*, although different in the start and intermediate states, while ending with the same ordering influences, for example, the state |2,1,3> of Table 4.5 (with "1" for LogP, "2" for POL, and "3" for E_{tot}, see Tables 4.3 and 4.4), provides weaker contribution to the recorder activity since two paths have to produce the same (final) effect in order it to be activated; this is nevertheless one remarkable mechanistic consequence of the present combined (algebraic or statistical) correlations with minimization (optimization) principle applied for the spectral path lengths through equations (4.6)–(4.10);

vii. The *alpha*, *beta*, and *gamma* paths can be easily identified for algebraic and statistical treatments in Tables 4.5 and 4.6 and there are accordingly marked; the degeneracy behavior is readily verified in Table 4.5 where the *alpha path is found as the only (non-degenerate) path* out of all possible ones. Of course, the same rationalization applies also for *alpha* path of Table 4.6, however displaying the trivial situation in which the absence of any degeneracy is recorded due to the restrained structural parameters considered for activity modeling since less available data for *Pimephales promelas* (*P.p.*) and *Vibrio fisheri* (*V.f.*) species in Table 4.2, according with the above specified Topliss-Costello rule.

Now, the inter-species analysis may be unfolded employing the paths of Tables 4.5 and 4.6; for achieving that, a preliminary search for minimum paths at the inter-species levels for each Mlog/Clog and algebraic/statistic computational frames should be done first. Note that for Daphnia magna (D.m.) species, although specific paths would be superfluous with the uni-parameter models considered in Tables 4.3 and 4.4 due the Topliss–Costello rule (since the limited data of Table 4.2), its presence on the inter-species grids of Figures 4.1–4.4 may be as well considered by means of the pseudo-path construction based on reconsidering the above (i) and (ii) minimum searching rules for models with single parameter dependency:

viii. Models with higher correlation/probability (either within statistic or algebraic approaches) will first enter molecular mechanism of toxicity through their considered structural parameter, that is, LogP, POL, and E_{tot} for the |1>, |2>, and |3> end-points, respectively.

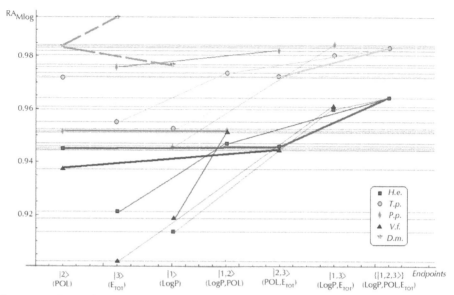

Figure 4.1. The *Hydractinia echinata* (*H.e.*), *Tetrahymena pyriformis* (*T.p.*), *Pimephales promelas* (*P.p.*), *Vibrio fisheri* (*V.f.*), and *Daphnia magna* (*D.m.*) inter-species SPECTRAL-SAR map modeling the molecular mechanisms for Mlog-algebraic toxicity paths of Tables 4.5 and 4.6 connecting the algebraic correlations of Table 4.3 across the ordered models of Table 4.7; the difference between species is made by the assignments of distinct icons, while *alpha, beta,* and *gamma* paths are differentiated by thickness decreasing of lines joining the same icons; the *D.m.* pseudo-path (interrupted line on map) is considered from the highest correlation model toward the lowest one in Table 4.3.

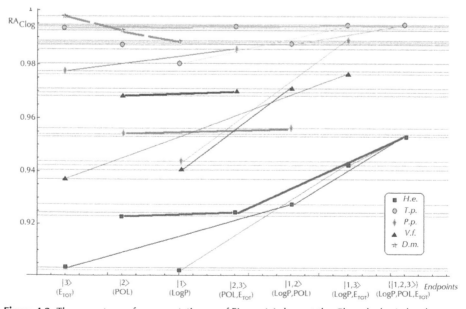

Figure 4.2. The same type of representation as of Figure 4.1, here at the Clog-algebraic level.

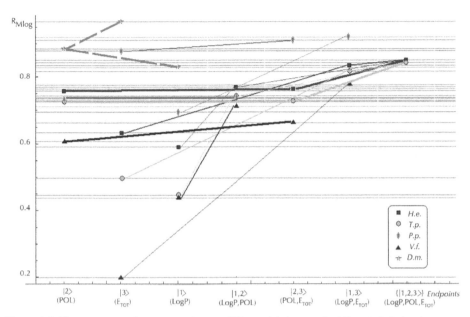

Figure 4.3. The same type of representation as of Figure 4.1, here at the Mlog-statistic level.

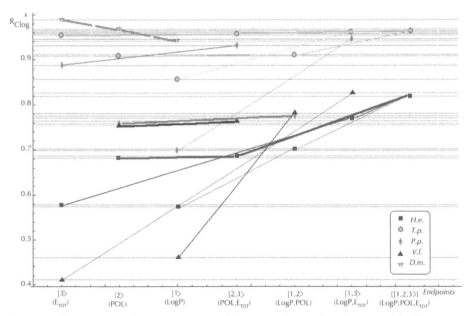

Figure 4.4. The same type of representation as of Figure 4.1, here at the Clog-statistic level.

Such a quest is performed in two steps: the computational scheme is primarily fixed, for example, the Mlog-algebraic one; then, among all Mlog-algebraic *alpha* paths for all species of Tables 4.5 and 4.6 the minimum is selected, that is, $\alpha_{P.p.}$ for the actual case.

Then, the same procedure is unfolded for the remaining *beta* and *gamma* paths within the fixed computational frame, that is, it will be repeated for each possible Mlog/Clog-algebraic/statistic combination. The results are summarized in Table 4.7 leading with the inter-species ordering of models to be considered for a mechanistic SPECTRAL-SAR analysis. As such, all possible inter- and intra-species influences are presented in Figures 4.1–4.4 emphasizing on primary (*alpha*), secondary (*beta*), and tertiary (*gamma*) paths of Tables 4.5 and 4.6 projected on the Mlog/Clog models for algebraic/statistic correlations of Tables 4.3 and 4.4, respectively.

Table 4.7. Synopsis of the inter-species minimum paths and the associated ordered endpoints for each of the Mlog/Clog-algebraic/statistic modes of computations abstracted from the Tables 4.5 and 4.6.

Computational Modes		Minimum Inter-species Paths			Ordered Endpoints
		alpha	*beta*	*gamma*	
Algebraic	Mlog	$\alpha_{P.p.}$	$\beta_{P.p.}$	$\gamma_{T.p.}$	
		$\|2{>}{\rightarrow}\|1{,}2{>}$	$\|3{>}{\rightarrow}\|2{,}3{>}$	$\|1{>}{\rightarrow}\|1{,}3{>}$	$\|2{>}{\rightarrow}\|3{>}{\rightarrow}\|1{>}{\rightarrow}\|1{,}2{>}{\rightarrow}\|2{,}3{>}{\rightarrow}\|1{,}3{>}{\rightarrow}\{\|1{,}2{,}3{>}\}$
	Clog	$\alpha_{T.p.}$	$\beta_{T.p.}$	$\gamma_{T.p.}$	
		$\|3{>}{\rightarrow}\|2{,}3{>}$	$\|2{>}{\rightarrow}\|1{,}2{>}$	$\|1{>}{\rightarrow}\|1{,}3{>}$	$\|3{>}{\rightarrow}\|2{>}{\rightarrow}\|1{>}{\rightarrow}\|2{,}3{>}{\rightarrow}\|1{,}2{>}{\rightarrow}\|1{,}3{>}{\rightarrow}\{\|1{,}2{,}3{>}\}$
Statistic	Mlog	$\alpha_{P.p.}$	$\beta_{P.p.}$	$\gamma_{P.p.}$	
		$\|2{>}{\rightarrow}\|1{,}2{>}$	$\|3{>}{\rightarrow}\|2{,}3{>}$	$\|1{>}{\rightarrow}\|1{,}3{>}$	$\|2{>}{\rightarrow}\|3{>}{\rightarrow}\|1{>}{\rightarrow}\|1{,}2{>}{\rightarrow}\|2{,}3{>}{\rightarrow}\|1{,}3{>}{\rightarrow}\{\|1{,}2{,}3{>}\}$
	Clog	$\alpha_{T.p.}$	$\beta_{T.p.}$	$\gamma_{T.p.}$	
		$\|3{>}{\rightarrow}\|2{,}3{>}$	$\|2{>}{\rightarrow}\|1{,}2{>}$	$\|1{>}{\rightarrow}\|1{,}3{>}$	$\|3{>}{\rightarrow}\|2{>}{\rightarrow}\|1{>}{\rightarrow}\|2{,}3{>}{\rightarrow}\|1{,}2{>}{\rightarrow}\|1{,}3{>}{\rightarrow}\{\|1{,}2{,}3{>}\}$

The inter-species diagrams reveal interesting features respecting both the correlation analysis and the inter-toxicity; as such, when is about either of algebraic or statistical treatment either Mlog- and Clog-inter-species ecotoxicity diagrams display the same endpoint ordering, as revealed by Table 4.7 and Figures 4.1 and 4.3 and 4.2 and 4.4, respectively. Beyond this, the algebraic approaches provide better systematic maps of inter-toxicity judged upon the *minimum distribution of crossing individual species' paths* (*alpha, beta,* or *gamma*), being this another realization of the least path principle—here at inter-species paths' level; for instance, the *H.e.* paths within algebraic framework RA_{Clog} of Figure 4.2 are clearly individuated as having no crossing toxicity with other species eventually submersed in the same ecological area, while the carried toxicity may be transmitted to *V.f.* species according with the statistical approach R_{Clog} of Figure 4.4.

On the other way, when comparing the measured (observed) results it is apparent that the species *H.e.* and *V.f.* are eco-toxically inter-connected and somehow independent from the *T.p.* and *P.p.* environmental response in RA_{Mlog} picture of Figure 4.1. Yet, a different situation is noted for the statistical R_{Mlog} analysis of Figure 4.3, according which *H.e.* species is highly mixed from a toxicological point of view with the species

T.p. and *P.p.*, but not with the *V.f.* one, either by means of first (*alpha*), second (*beta*), or third (*gamma*) toxicity paths.

Finally, the species *D.m.* is predicted to strongly interact (crosses at the *alpha* paths' level) with the species *T.p.* on both algebraic RA_{Clog} and statistical R_{Clog} frameworks of Figures 4.2 and 4.4 due to POL and LogP parameters specific influence—identified on the grid region of their path crossings, respectively. Such a situation is no longer valid when Mlog values are modeled, since the algebraic RA_{Mlog} approach predicts moderate inter-toxicity influence (through *alpha–beta* crossing paths due E_{tot} or steric influence) (see Figure 4.1), in contrast with no recorded interaction within the statistical R_{Mlog} analysis (see Figure 4.3).

Therefore, the molecular mechanistic models of toxicity may be proposed in four variants: based on algebraic (RA) or statistic (R) correlation of either measured (Mlog) or by *ESIP* computed (Clog) toxicities.

The difference between the algebraic and statistical approaches relays on their inner definition: while, for a data sample, the statistical framework quantifies the dispersion respecting the data average (the data mean), the algebraic picture accounts for the dispersion of the extremes (the *N*-dimensional Euclidian lengths of the data rows); from this conceptual difference, although both assess the same confined realm between zero and one in probability realization, the algebraic correlation records closer values near to the certainty for models classified as with high or even moderate statistical correlation values [7], being thus more suited for *least path principle* applications, as also proven by the current study.

In other words, if one is interested in the sample data behavior merely from its "length" (the norm) than from its "average dispersion" side, the algebraic way should be chosen as the main correlation framework, while keeping the statistical counterpart available for comparison purpose. This seems to be the case of ecotoxicological studies when the intensity (the "length or the amplitude") of action for each sample's endpoint may be important [8].

On the other hand, the difference between the measured and computed values for ecotoxicological activities relays on the way the *ESIP* model is build from the available database, that is, by collecting the measured molecular-upon-species values and then appropriately redistributing them among various molecular fragments and groups of actual interest. Nevertheless, a practical discussion on how fine the actual *ESIP* data accommodates with the correlated structural data is addressed next.

DISCUSSION OF ESIP

Table 4.2 presents the 28 tested combinations of *H.e.* with other organisms. In the case of derivatives numbers 1–25, the estimations of the calculated (C) values have been possible through use of the *ESIP*'s parameters distinctive for molecular substructures and specifically for every test-system. The file (structure + *ESIP* algorithm) of the mentioned derivatives offers the possibility of the analysis of different structure-reactivity relations through mentioned organisms we follow.

In the *H.e.* case, a specifically marine environment organism, pure water has the least toxicity and has the least values for structural parameters. The appearance of the hydrocarbonated chain leads to increasing molecular toxicity simultaneously with the increased values of the structural parameters (logP and POL) in the case of alcohols (nos. 2–4, and 6) and in the case of the phenols (nos. 14, 15, and 21) too. Compound number 21 has the highest toxicity, probably due to geometry of the hydrocarbon radical situated in opposite *para-* position for phenolic hydroxyl and this one proximity on aromatic nucleus [27]; It is worth observing that the steric impediments limit produces such increase in the case of 2,6-diisopropylphenol. Instead, 1,2,3-propanetriol possess three OH groups, leading with persistent hydrophilic character, while the molecule has a diminished toxicity according to the logP value.

In the aromatic series (molecules nos. 16, 18, and 20) the structural parameters have closer values but the toxicities are more elevated as a result of the possibilities of extended electronic conjugation. The existence of two identical hydroxyl groups (see molecule no. 18), a highly symmetrical and flat molecule, as well as the absence of sterical hindrances, are considered to be the premises of an extended p-p conjugation (the possibility of conjugation between the non-bonded p electrons of oxygen and the p electrons of aromatic center) according to a push–pull electronic mechanism: an OH group is electron donating and becomes positively-charged, and the second one, an electron accepting group becomes negatively-charged. This phenomenon, which is probably alternant and permanent even in the absence of a reaction partner, induces a strong hydrogen bond donor character. Unexpected seems the toxicity of 1,2,3-trihydroxibenzene (molecule no. 20), essentially identical with that of 1,2-dihydroxybenzene, although its logP value is diminished; this maybe happens due the push–pull mechanism of polyhidroxylic phenols [6].

In the case of 1,4-dihydroxybenzene, the conjugation is diminished through the inclusion of a *t*butyl radical (no. 19) and increased steric impediment at the phenolic hydroxyl level. However, the toxicity significantly decreases by three orders by replacing one—OH group with a methyl (no. 14), methoxy (no. 17), chloro (no. 22), or amino (no. 24) moiety though the logP parameter changes significantly. The situation according to which the toxicity values of the mentioned derivatives span a narrow domain of about 0.5 logarithm units, relays on the existence of certain stereo-electronic balance similarity in the case of aromatic derivatives inferior substituted, as 4-toluidine (no. 11) and 1,2-dichlorobenzene (no. 12), in agreement with other (unpublished) series of derivatives.

The toxicity dependence on the logP parameter is informative also for derivatives with pyridine nitrogen (no. 8–10), quinonic ones (no. 26 and 27), and 2,4,6-trinitrophenol (no. 25), which although having the most diminished logP value, displays however identical efficacy with monosubstituted phenol derivatives.

An example of the influence of functional groups is illustrated by the 1,10-diaminodecane (no. 7) molecule with a smaller toxicity than expected according to its hydrocarbonated chain, though the POL value is great. According with the experimental results [5], the chain with 8–10 C probably represents the hydrocarbon interface where the lipophilicity is manifested. Yet, the diffusion of the molecules with high number of the C atoms through the cell membrane is however "hindered" and the percentage of the crossing molecules is diminished [28].

The presence of the 4-methoxyazobenzene (no. 28) in Table 4.2 illustrates that as profound structural changes resulted new reactions mechanisms appear, emphasizing the availability offered by the test-system with *H.e.* to analyze very different derivatives. This is the case of the azo-function which, by means of enzymatic reduction, leads to the stoichiometric appearance of amines, though the toxicity of the mixed combination (the most frequent case in the environment) represents a fruitful and significant investigation direction.

The *ESIP*-Köln model provides, although not in all cases, the possibility to appreciate to a great extent the efficiency with which the real or measured M value agrees with the theoretically calculated (C) counterparts. In other words, if the calculated value (C) stands above those measured (M), for instance, the difference can be assigned to the lipophilicity character represented here by logP, along some electronic POL or steric Etot influences.

However, the *ESIP* values have been determined to bend down on real/measured values through inclusion of the effects relating specific parameters of molecular structures. This is confirmed since the measured $Mlog/MRC_{50}$ and calculated $Clog/MRC_{50}$ values are in concordance with the numerical structural parameter counterparts in the Table 4.2. This also indicates that the individual structural parameters or their combinations are specific and the organism *H.e.* can be successful employed as a suitable test-system for further toxicity determinations.

CONCLUSION

There is already wider recognition of the problem posed by the ever growing number of available chemicals with no tested toxicity in junction with the increased costs and limited time available for testing before entering mainstream production or they are dispersed into the environment. Therefore, the demand for developing *in silico* tools for providing the associated computed activities from benchmark measurements and individuated molecular fragment toxicities naturally appears; such studies should provide correlation paths regarding how the given toxicants may act on various cells or species. In moving toward, such complex computational techniques for species and inter-species toxicity assignment the present work combines the Köln-*ESIP* and Timişoara-SPECTRAL-SAR models in a unified computational activity-correlation framework.

The Köln model for estimation of a compound toxicity is based on the experimental measurement expressing the direct action of chemicals on the *H.e.* organism so that the structural influence parameters are reflected by the degree of metamorphosis itself. As such, the calculation of the structural parameters is absolute necessary for correctly evaluation and interpretation of the evolution of M(easured) and C(omputed) values.

The present work evidences relatively simple rules in respect to relationships with structure and reactivity: the efficacy of aliphatic alcohols increases with the number of the C atoms (a phenomenon characterized through the structural parameter logP) and diminishes with the appearance of new alcoholic groups (a phenomenon widely reflected through POL and E_{tot}); the influence of the amino-group for aliphatic amines is comparatively predominant to the relative extent of the hydrocarbon chain [5]; the influence of the first methyl in the phenol case is negligible; the steric influence of isopropyl-radicals on the phenolic active center for 2,6-derivative is stronger (by increased POL value) as in the case of the *t*-butyl substituent (while logP and POL values are diminished), and so forth.

In principle, the toxicity represents the synergetic effect of the three structural influences: hydrophobicity, electrostatic, and steric molecular control on receptor binding. The efficiency difference through derivatives with closer parameters as those with 1,2-, 1,4-, and 1,2,3-hydroxy groups can be interpreted through a particular electronic mechanism [6]; on the other hand, the toxicity efficiency difference through 1,4-dihydroxybenzene (electronic) and 4-(3′,5′-dimethyl-3′-heptyl)phenol (steric) is well reflected through computed parameter's numeric values.

Finally, the Timişoara SPECTRAL-SAR analysis [21] offers the correlation models and the end-points' paths either for *H.e.* species as well for other four different organisms, with which the toxicity may be inter-changed by means of molecular structural mechanisms of action induced by certain common (or under testing) chemicals.

Besides the fact the SPECTRAL-SAR algorithm was previously proven as being superior to the fashioned statistical approach in solving the paradoxical dichotomies various statistical indices produce when considered together [24], it advances a reliable method of identifying which structural molecular parameter is more influential across multiple possible paths of activation of a bio- or ecotoxicological response, thus furnishing a useful computational mechanistic molecular method in QSAR studies [29]. Then, when combined with *ESIP* algorithm a complex inter-molecular/inter-species toxicological transfer picture is provided.

However, the correlation maps depend on the algebraic or statistical way of modeling the action, that is, by assuming the chemical–biological interaction driven by the intensity norm (relating algebraic vectorial picture) or average (relating statistical dispersive picture) in ligand-receptor specific binding. At this point the algebraic versus statistic issue remains open for further investigations by comparative single- and inter-species activities.

APPENDIX: GENERALIZED EUCLIDIAN CORRELATION FACTOR

Vectorial Scalar Product, Norms, and Cauchy-Schwarz Inequality

Given two vectors:

$$|u\rangle = |u_1, u_2, ..., u_n\rangle, \quad |v\rangle = |v_1, v_2, ..., v_n\rangle, \tag{4.A1}$$

their scalar (or dot) product writes as:

$$\langle u|v\rangle \overset{def}{=} \sum_{i=1}^{n} u_i v_i = u_1 v_1 + u_2 v_2 + ... + u_n v_n. \tag{4.A2}$$

Since the *self* scalar product looks like:

$$\langle u|u\rangle = \sum_{i=1}^{n} u_i^2 \tag{4.A3}$$

one may introduce the so called norm (or length) of the vector by definition:

$$\||u\rangle\| = \sqrt{\langle u|u\rangle} = \sqrt{\sum_{i=1}^{n} u_i^2} \ . \tag{4.A4}$$

The length property of the vectorial norm may be easily visualized through computing the modulus of an arbitrary 3D vector, say $|r\rangle = |u_1, u_2, u_3\rangle$:

$$|\vec{r}| = \sqrt{u_1^2 + u_2^2 + u_3^2} = \sqrt{\langle u_1, u_2, u_3 | u_1, u_2, u_3 \rangle} = \sqrt{\langle r|r\rangle} = \||r\rangle\|. \tag{4.A5}$$

Consequently, the distance between two vectors may be written in terms of their difference norm as:

$$d(|u\rangle, |v\rangle) = \||u\rangle - |v\rangle\| = \||u - v\rangle\| = \sqrt{\langle u - v|u - v\rangle} = \sqrt{\sum_{i=1}^{n}(u_i - v_i)^2} \tag{4.A6}$$

From equation (4.A6), but also from the fact that the self-scalar product is positively defined, see equation (4.A4), the distributivity and commutativity properties of scalar product may be employed for any real parameter, $t \in \Re$, toward equivalent expressions:

$$\begin{aligned}\langle u - tv|u - tv\rangle &\geq 0, \ t \in \Re \\ \Leftrightarrow \left((\langle u| - \langle v|t)(|u\rangle - t|v\rangle)\right) &\geq 0 \\ \Leftrightarrow \langle v|v\rangle t^2 - 2\langle u|v\rangle t + \langle u|u\rangle &\geq 0\end{aligned} \tag{4.A7}$$

The last inequality says that the left sided second order equation has no solution or has two equal solutions; such condition is fulfilled when its discriminator is less or equal with zero, respectively, leading with the famous *Cauchy–Schwartz inequality*:

$$\langle u|v\rangle^2 \leq \langle u|u\rangle\langle v|v\rangle, \tag{4.A8}$$

which may be rewritten as:

$$\sum_{i=1}^{n} u_i v_i \leq \sqrt{\sum_{i=1}^{n} u_i^2} \sqrt{\sum_{i=1}^{n} v_i^2} \tag{4.A9}$$

or in more formal way as:

$$|\langle u|v\rangle| \leq \||u\rangle\| \cdot \||v\rangle\|. \tag{4.A10}$$

The Cauchy–Schwartz inequality is successfully used in probability theory, variance theory, and correlation factors. An actual application is in following given as well.

Algebraic Correlation Factor
One starts with the simple connection between the observed, predicted, and error vectors of equation (4.1a), however specialized on their individual elements:

$$Y_{i-OBS} = Y_{i-PRED} + pe_i \tag{4.A11}$$

where "*pe*" here stays as the abbreviation for "prediction error."

Then, while squaring relation (4.A11):

$$Y_{i-OBS}^2 = Y_{i-PRED}^2 + pe_i^2 + 2Y_{i-PRED} \cdot pe_i, \tag{4.A12}$$

and summing for all working *N*-molecules (of Table 4.1):

$$\sum_{i=1}^{N} Y_{i-OBS}^2 = \sum_{i=1}^{N} Y_{i-PRED}^2 + \sum_{i=1}^{N} pe_i^2 + 2\sum_{i=1}^{N} Y_{i-PRED} \cdot pe_i, \tag{4.A13}$$

the last relation simplifies to:

$$\sum_{i=1}^{N} Y_{i-OBS}^2 = \sum_{i=1}^{N} Y_{i-PRED}^2 + \sum_{i=1}^{N} pe_i^2 \tag{4.A14}$$

based on applying of scalar product definition (4.A2) and of prediction error orthogonalization condition (4.1c) for the last term of (4.A13):

$$\sum_{i=1}^{N} Y_{i-PRED} \cdot pe_i = \langle Y_{PRED} | pe \rangle = 0. \tag{4.A15}$$

Now, substituting the prediction error values of (4.A11) in (4.A14) one equivalently yields:

$$\sum_{i=1}^{N} Y_{i-OBS}^2 = \sum_{i=1}^{N} Y_{i-PRED}^2 + \sum_{i=1}^{N} (Y_{i-OBS} - Y_{i-PRED})^2$$

$$\Leftrightarrow \sum_{i=1}^{N} Y_{i-PRED}^2 = \sum_{i=1}^{N} Y_{i-OBS} \cdot Y_{i-PRED} \tag{4.A16}$$

which further rewrites, recalling the norm and scalar product definitions of equations (4.A4) and (4.A2), respectively, as:

$$\left\| Y_{PRED} \right\rangle \right\|^2 = \langle Y_{OBS} | Y_{PRED} \rangle. \tag{4.A17}$$

Finally, the Cauchy–Schwarz form (4.A10) is employed on the right side term of (4.A17), noting that the observed and predicted activities are of the same nature for a given molecule—that is, either both positive or both negative—thus providing their scalar product as positively defined; yet this is certified also by the relation (4.A17) viewed as a result *per se*; nevertheless, through considering Cauchy–Schwarz prescription, the relation (4.A17) immediately transforms into inequality:

$$\left\| Y_{PRED} \right\rangle \right\|^2 \leq \left\| Y_{OBS} \right\rangle \right\| \cdot \left\| Y_{PRED} \right\rangle \right\| \tag{4.A18}$$

leaving with the predicted-observed norms' hierarchy:

$$\left\| Y_{PRED} \right\rangle \right\| \leq \left\| Y_{OBS} \right\rangle \right\|. \tag{4.A19}$$

Relation (4.A19) guarantees the *consistent probability definition* for the introduced *algebraic correlation factor* of equation (4.8):

$$r_{ALGEBRAIC} = \frac{\left\| \left| Y_{PRED} \right\rangle \right\|}{\left\| \left| Y_{OBS} \right\rangle \right\|} \equiv RA \le 1 . \tag{4.A20}$$

KEYWORDS

- **Correlation map**
- *ESIP* **Köln model**
- **SPECTRAL-SAR algorithm**
- **Topliss–Costello rule**

Chapter 5

Turning SPECTRAL-SAR into 3D-QSAR Analysis: Application on Proton-pump Inhibitory Activity

Mihai V. Putz, Corina Duda-Seiman, Daniel M. Duda-Seiman, and Ana-Maria Putz

INTRODUCTION

Within the SPECTRAL-SAR structure-activity correlation environment, the vectorial algebraic version of the consecrated QSAR analysis, the minimum spectral endpoint-connected paths hierarchy is employed to provide the 3D molecular structure that closely resemble the maximum observed chemical–biological activity across a given series of compounds. The algebraic spectral proposition asserting that the algebraic correlation factors always lies above the dispersion based statistically one for the same series of predicted-observed data was here formulated and demonstrated. As practical illustration of the introduced 3D-SPECTRAL-SAR method the special case of proton-pump inhibitors (PPIs), blocking the acid secretion from parietal cells in the stomach, here belonging to the 4-Indolyl, 2-guanidinothiazole class of derivatives, was exposed and the optimized-by-the-spectral mechanism molecular structure revealed.

Inhibition of gastric acid secretion represents the most attractive target in treating acid-related diseases, like duodenal and gastric ulcer, gastroesophageal reflux disease (GERD), Zollinger–Ellison syndrome, and Barrett's esophagus. By these means, proton-pump inhibitors became first-line pharmacological agents in treating the above-mentioned pathological conditions [1]. The therapeutic regimen that combines two antimicrobials (clarithromycin and amoxicillin or metronidazole) with a PPI is well-established in the eradication management of *Helicobacter pylori* infection [2].

In clinical practice PPIs are used with chemical structures of substituted benzimidazoles [3]. However, there are several PPIs that have been synthesized, some of them with other chemical structures: tenatoprazole (marked as TU-199) consists of one imidazopyridine ring connected to a pyridine ring by a sulfinylmethyl chain [2] (see Figure 5.1).

Gastric parietal cells are responsible for acid secretion, controlled through several food-stimulated and neuroendocrine pathways mediated by gastrin, histamine, pituitary adenylate cyclase-activating peptide, and acetylcholine [4]. PPIs target the final effector in the acid secretion pathway, namely the gastric H^+/K^+-ATPase [4]. In acidic environment, PPIs as prodrugs are converted to active sulfenamide form; then, the active drug forms a covalent bond to cysteine residues of active proton pumps producing an almost complete inhibition of gastric acid secretion which lasts until new proton pump molecules are synthesized [5].

(a)

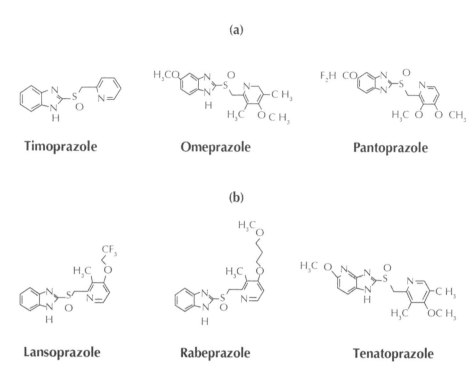

Timoprazole   Omeprazole   Pantoprazole

(b)

Lansoprazole   Rabeprazole   Tenatoprazole

Figure 5.1. Proton pump types of inhibitors based on benzimidazole and imidazopyridine rings in (a) and (b) respectively [1, 2].

There are irreversible and reversible gastric PPIs [6]. Irreversible gastric PPIs are structurally mainly characterized by the presence of the substituted pyridine ring, of the substituted benzimidazole ring, and of the methylsulfinyl linking group [6]. They are highly effective in gastric acid secretion inhibition leading in prolonged administration to extreme acid suppression and achlorohydria. Their most severe side-effects are hypergastrinemia, gastric polips, and gastric carcinoma. In order to avoid these side-effects, several reversible PPIs are under development, often referred as acid pump antagonists, but none is yet marketed [6].

In the effort to develop new effective drugs with gastric anti-secretor activity, thiazolidines show potential to inhibit the gastric H⁺/K⁺-ATPase [7].

Nevertheless, among many competitive-inhibitory pathways of H⁺/K⁺-ATPase proton pump, while the main chemical–physical process involved regards the PPIs diffusion through parietal cells in the stomach, the steps concerned in the signal transduction have not yet been elucidated. In this context, the present study addresses the employment of the recent proposed SPECTRAL-SAR algorithm [8-13] to predict the most active molecular structure among a given series of PPI compounds by means of pre-designed activity paths.

BASICS OF THE SPECTRAL-SAR METHOD

Basically, the applied SPECTRAL-SAR algorithm can be summarized by means of the so called SPECTRAL-SAR operator defined as a collection of successive operations,

$$
\hat{O}_{S-SAR} : \begin{cases} Det\left(|Y\rangle, |X_0\rangle, |X_1\rangle, ..., |X_M\rangle\right) = 0, \|Y\|, r_{S-SAR}^{ALGEBRAIC}, \\ \delta\left[A_{(|\cdot|r)}, B_{(|\cdot|r)}\right] = 0, \langle\alpha, \beta, \gamma, ...\rangle, \quad \begin{matrix} A, B : ENDPOINTS \\ \alpha, \beta, \gamma, ... : SPECTRAL\ PATHS \end{matrix} \end{cases}, \quad (5.1)
$$

employing the N-structural molecular input data, as the vectors $|X_0\rangle = |1_1...1_N\rangle$, $|X_i\rangle$, $i=1,...M$, with the predicted endpoints $|Y\rangle$, see Table 5.1, *via* the scalar product-based spectral norm $\||Y\rangle\|$, by using algebraic correlation factor $r_{S-SAR}^{ALGEBRAIC}$ as the ratio of calculated to the measured spectral norm, until the first minimum paths across the spectral norms-correlation factors are attained between the specific endpoints [10].

Unfolding the above operator, the classical QSAR equation is here replaced by its vectorial version,

$$
|Y\rangle = b_0|X_0\rangle + b_1|X_1\rangle + ... + b_k|X_k\rangle + ... + b_M|X_M\rangle, \quad (5.2)
$$

while the analytical counterpart is provided by the associated determinant [8, 10]

$$
\begin{vmatrix} |Y\rangle & \omega_0 & \omega_1 & \cdots & \omega_k & \cdots & \omega_M \\ |X_0\rangle & 1 & 0 & \cdots & 0 & \cdots & 0 \\ |X_1\rangle & r_0^1 & 1 & \cdots & 0 & \cdots & 0 \\ \vdots & \vdots & \vdots & \vdots & \vdots & & \vdots \\ |X_k\rangle & r_0^k & r_1^k & \cdots & 1 & \cdots & 0 \\ \vdots & \vdots & \vdots & \vdots & \vdots & & \vdots \\ |X_M\rangle & r_0^M & r_1^M & \cdots & r_k^M & \cdots & 1 \end{vmatrix} = 0, \quad (5.3)
$$

where the components are calculated with the recipes

$$
r_i^k = \frac{\langle X_k|\Omega_i\rangle}{\langle\Omega_i|\Omega_i\rangle}, \omega_k = \frac{\langle\Omega_k|Y\rangle}{\langle\Omega_k|\Omega_k\rangle}, k = \overline{0, M}, \quad (5.4)
$$

by means of the Gram–Schmidt algorithm:

$$
|\Omega_0\rangle = |X_0\rangle; |\Omega_k\rangle = |X_k\rangle - \sum_{i=0}^{k-1} r_i^k|\Omega_i\rangle. \quad (5.5)
$$

The advantage of the SPECTRAL-SAR method stands in the following points: it allows direct correlation equation without appealing to the computational methods of solving large systems with differential equations; it provides a transparent, algebraically direct method; it employs the orthogonality constraint of the correlated data;

treated the data as vectors thus providing an equation representative not only to individual molecular actions but also to an entire set of chemicals action on certain biological system.

Table 5.1. The SPECTRAL (vectorial)—SAR generic table of descriptors.

Activity	Structural Predictor Variables					
$\lvert Y \rangle$	$\lvert X_0 \rangle$	$\lvert X_1 \rangle$	$\cdots$	$\lvert X_k \rangle$	$\cdots$	$\lvert X_M \rangle$
y_1	1	x_{11}	$\cdots$	x_{1k}	$\cdots$	x_{1M}
y_2	1	x_{21}	$\cdots$	x_{2k}	$\cdots$	x_{2M}
$\vdots$	$\vdots$	$\vdots$	$\vdots$	$\vdots$	$\vdots$	$\vdots$
y_N	1	x_{N1}	$\cdots$	x_{Nk}	$\cdots$	x_{NM}

Beside these the SPECTRAL-SAR approach of structure-activity correlation opens new perspectives of analytical analysis of chemical–biological interaction. Actually, since the vectorial frame of computation the so called vectorial norm of a certain model (endpoint) either measured or computed may be introduced by the consecrated N-generalized definition of the norm-scalar product rule:

$$\left\lVert Y_{OBS/PRED} \right\rangle \right\rVert = \sqrt{\sum_{i=1}^{N} y_{i-OBS/PRED}^2} \ . \tag{5.6}$$

Next, the vectorial norm gives the opportunity in introducing the so called algebraic correlation factor measuring the relative "intensity" or "amplitude" of the predicted to measured norm activities (see Appendix 4.6.A2 of the Chapter 4 of the present volume):

$$r_{S-SAR}^{ALGEBRAIC} = \frac{\left\lVert Y_{PRED} \right\rangle \right\rVert}{\left\lVert Y_{OBS} \right\rangle \right\rVert} \leq 1. \tag{5.7}$$

However, worth noting that the algebraic correlation factor gives systematic higher values respecting the assumed statistical one, computed on the statistics of the data dispersion through the expression

$$r_{QSAR}^{STATISTIC} = \sqrt{1 - \frac{\sum_i (y_{i-OBS} - y_{i-PRED})^2}{\sum_i (y_{i-OBS} - \bar{y}_{OBS})^2}}, \tag{5.8}$$

for each of the studied cases so far [14]. Such behavior is susceptible to hold in general as it will be next exposed. Actually, we will prove the next proposition:

Banater Ansatz on the algebraic SPECTRAL-SAR correlation [15]: For any QSAR analysis, once considering the measured/observed and computed/predicted activity data as the vectors $\left| Y_{OBS} \right\rangle$ and $\left| Y_{PRED} \right\rangle$ with the associate norms through the scalar products—see 4.6.A2 of the Chapter 4 of the present volume, the algebraic norm order

$$\left\| Y_{PRED} \right\rangle \right\| \leq \left\| Y_{OBS} \right\rangle \right\| \tag{5.9a}$$

valid in defining the algebraic correlation factor (5.7), sets also the hierarchy at the levels of correlations factors

$$r_{S-SAR}^{ALGEBRAIC} \geq r_{QSAR}^{STATISTIC} \tag{5.9b}$$

in a sense that the algebraic one of always exceed the standard correlation factor (5.8).

Proof: By straight algebraic translation, the condition (5.9b) with equations (5.7) and (5.8) in terms of scalar products, it first rewrites as:

$$\frac{\left\langle Y_{PRED} \middle| Y_{PRED} \right\rangle}{\left\langle Y_{OBS} \middle| Y_{OBS} \right\rangle} \geq 1 - \frac{\left\langle Y_{OBS} - Y_{PRED} \middle| Y_{OBS} - Y_{PRED} \right\rangle}{\left\langle Y_{OBS} - \bar{Y}_{OBS} \middle| Y_{OBS} - \bar{Y}_{OBS} \right\rangle}, \tag{5.10}$$

where we have introduced the averaged observed activity

$$\bar{Y}_{OBS} = \frac{1}{N} \sum_{i=1}^{N} y_{i-OBS}, \tag{5.11a}$$

and its associate N-dimensional vector (state in Hilbert space):

$$\left| \bar{Y}_{OBS} \right\rangle = \left(\frac{1}{N} \sum_{i=1}^{N} y_{i-OBS} \right) \left| 1\,1 \ldots 1_N \right\rangle. \tag{5.11b}$$

Note that the inequality (5.10) becomes equality in the case of perfect identity between observed and predicted activity values, that is perfect correlation, the case in which the second term of the right hand side vanishes while that of the left hand side become unity. For all other non-perfect correlations strict inequality holds and this will be considered in next, for the equivalent expression

$$\begin{aligned}
&\left\langle Y_{PRED} \middle| Y_{PRED} \right\rangle \left\langle Y_{OBS} - \bar{Y}_{OBS} \middle| Y_{OBS} - \bar{Y}_{OBS} \right\rangle \\
&> \left\langle Y_{OBS} \middle| Y_{OBS} \right\rangle \left[\left\langle Y_{OBS} - \bar{Y}_{OBS} \middle| Y_{OBS} - \bar{Y}_{OBS} \right\rangle - \left\langle Y_{OBS} - Y_{PRED} \middle| Y_{OBS} - Y_{PRED} \right\rangle \right],
\end{aligned} \tag{5.12}$$

which may be further rearranged as

$$\begin{aligned}
&\left[\left\langle Y_{PRED} \middle| Y_{PRED} \right\rangle - \left\langle Y_{OBS} \middle| Y_{OBS} \right\rangle \right] \left[\left\langle Y_{OBS} \middle| Y_{OBS} \right\rangle - 2\left\langle Y_{OBS} \middle| \bar{Y}_{OBS} \right\rangle + \left\langle \bar{Y}_{OBS} \middle| \bar{Y}_{OBS} \right\rangle \right] \\
&> \left\langle Y_{OBS} \middle| Y_{OBS} \right\rangle \left[\left\langle Y_{OBS} - \bar{Y}_{OBS} \middle| Y_{OBS} - \bar{Y}_{OBS} \right\rangle - \left\langle Y_{OBS} - Y_{PRED} \middle| Y_{OBS} - Y_{PRED} \right\rangle \right].
\end{aligned} \tag{5.13}$$

At this point, after obvious simplifications and factorization may easily recognize and employ both the identities (4.A17) and (4.A19), specific to algebraic correlation,

$$2\langle Y_{PRED}|Y_{PRED}\rangle\langle Y_{OBS}|Y_{OBS}\rangle - 2\langle Y_{OBS}|Y_{OBS}\rangle\underbrace{\frac{\langle Y_{OBS}|Y_{PRED}\rangle}{\langle Y_{PRED}|Y_{PRED}\rangle}}$$

$$+\underbrace{\left[\langle Y_{OBS}|Y_{OBS}\rangle - \langle Y_{PRED}|Y_{PRED}\rangle\right]}_{\geq 0}\left[2\langle Y_{OBS}|\bar{Y}_{OBS}\rangle - \langle\bar{Y}_{OBS}|\bar{Y}_{OBS}\rangle\right] > 0 \qquad (5.14)$$

to obtain the simplified expression

$$2\langle Y_{OBS}|\bar{Y}_{OBS}\rangle > \langle\bar{Y}_{OBS}|\bar{Y}_{OBS}\rangle \qquad (5.15)$$

that finally is analytically explicated with the aid of introduced vector (5.11b) of the average activity to the unfolded scalar ordered products

$$2\sum_{i=1}^{N}\left(y_{i-OBS}\frac{1}{N}\sum_{i=1}^{N}y_{i-OBS}\right) > \sum_{i=1}^{N}\left(\frac{1}{N}\sum_{i=1}^{N}y_{i-OBS}\right)\left(\frac{1}{N}\sum_{i=1}^{N}y_{i-OBS}\right) \qquad (5.16)$$

leaving with the equivalent strict inequality

$$\frac{2}{N}\left(\sum_{i=1}^{N}y_{i-OBS}\right)^2 > \frac{1}{N}\left(\sum_{i=1}^{N}y_{i-OBS}\right)^2 \qquad (5.17)$$

fully satisfied by the natural ordering as $2 > 1$. Therefore, there was proved both the (qualitative) simplicity and the (quantitative) superiority of algebraic correlation factor.

As a note worth pointing that the hypothesis (5.9a), beside being proved in Appendix 4.6.A2 of Chapter 4 of the present volume, it seems quite reasonably since it measures the vectorial "lengths" of the measured and predicted intensities of the chemical–biological interactions; moreover, this is sustained also by the final endpoint status carried by the observed activity vector; in fact any correlation analysis likes to find the proper structural combination factors, and therefore also the mechanisms, able to predict (therefore modeling) the observed endpoint as closely as possible. In this context, the vectorial picture assures the intuitive phenomenology in which the biological activity and its models are represented by vectors and of their lengths (norms) with the observed activity being the highest one among all models in the set of predicted mechanisms.

This way the SPECTRAL-SAR algorithm produces a correlation factor, the algebraic one (5.7), that provides a better correlation respecting the statistical classical approach; this contribution however opens an epistemological discussion through the question: which correlation is the real one? Here we only stipulate that the algebraic and statistical pictures are qualitatively different since the one is based on dispersion while the other on vectorial length of predicted against the observed biological activities, respectively. In short, the SPECTRAL-SAR analysis employs the idea of "amplitude" or "intensity" or "length" of chemical–biological interaction and bonding. In this context, appears also the idea of introducing of the least path principle that selects the optimal (shortest) paths among the computationally tested models toward the measured endpoint:

$$\delta[A,B] = 0; \quad A,B: ENDPOINTS .$$ (5.18)

Analytically, the paths are computed via the Euclidian distances between each consecutive two points of a certain spectral path of models in the abstract space of vectorial norms-correlation (recommended algebraically but being available also for statistically) factors:

$$[A,B] = \sqrt{\left(\left\|\left|Y^B\right\rangle\right\| - \left\|\left|Y^A\right\rangle\right\|\right)^2 + \left(r_B^{\substack{STATISTIC/ \\ ALGEBRAIC}} - r_A^{\substack{STATISTIC/ \\ ALGEBRAIC}}\right)^2} .$$ (5.19)

The optimized spectral paths out of all set of possible paths that connect the endpoints belonging to different modeling correlation classes, deliver the spectral paths hierarchies predicted; these are used to build a mechanistic picture of the molecular (ligand or effector) interaction with the substrate (organism or receptor) host.

The remaining issue is to use the SPECTRAL-SAR methodology in order to predict the 3D configuration of the effector–receptor binding through the predicted endpoints, their norms and algebraic factors, and minimum spectral paths targeting the minimum residuum of the predicted respecting observed chemical–biological activity. The first attempt of this goal is in the next section unfolded on special case of H^+K^+-ATPase inhibitory activity.

SPECTRAL-SAR RESULTS FOR ANTI-ULCER ACTIVITY

Application of the above SPECTRAL-SAR methodology is here exposed on a selected series of 4-indolyl, 2-guanidinothiazole derivatives in the Table 5.2 with the general structure of Figure 5.2.

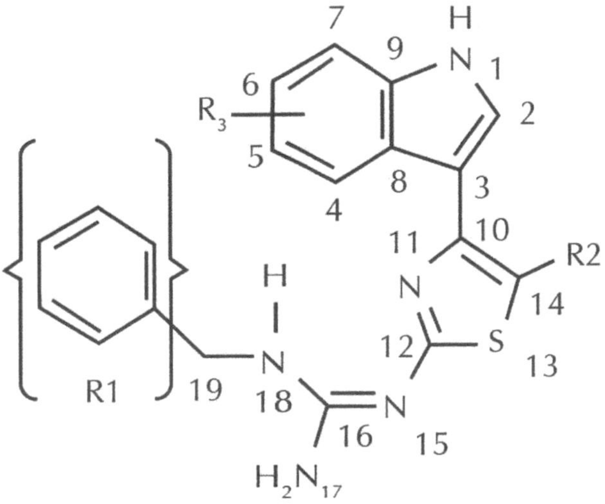

Figure 5.2. General scheme of 4-indolyl, 2-guanidinothiazole derivatives.

Table 5.2. Particular 4-indolyl, 2-guanidinothiazole derivatives with the associate structural descriptors (hydrophobicity, polarizability, and total energy), computed with the help of HyperChem [16], in modeling the H^+K^+-ATPase inhibitory activities [7].

No.	Name	R1	R2	R3	$|Y\rangle{=}A{=}$ $Log(1/IC50)$	$|X_1\rangle{=}$ $LogP$	$|X_2\rangle{=}$ $POL[Å^3]$	$|X_3\rangle{=}E_{TOT}$ $[kcal/mol]$
1	4-indolyl-2-methylguanidinothiazole	H	H	H	−2.1972	−1.0	30.27	−14.9839
2	4-indolyl-2-methylheptyl-guanidinothiazole	C_7H_7	H	H	−0.4055	−0.51	41.76	−29.4849
3	5-methyl-4-indolyl-2-methylguanidinothiazole	H	Me	H	−2.0281	−0.98	32.10	−27.3296
4	4-(5-methoxyindolyl)-2-methylguanidinothiazole	H	H	5-OMe	−2.9444	−2.00	32.74	−19.6745
5	4-(5-methoxyindolyl)-2-methylheptylguanidinothiazole	C_7H_7	H	5-OMe	−1.4586	−0.48	44.24	−35.2272
6	4-(5-benzyloxyindolyl)-2-methylguanidinothiazole	H	H	5-OC_7H_7	−0.1823	−0.51	42.40	−34.1524
7	4-(5-benzyloxyindolyl)-2- methylheptyl-guanidinothiazole	C_7H_7	H	5-OC_7H_7	−0.5878	1.01	53.90	−44.7714
8	4-(2-methylindolyl)-2-methylguanidinothiazole	H	H	2-Me	−2.1972	−0.98	32.10	−21.9728
9	4-(2-methylindolyl)-2-methylheptylguanidinothiazole	C_7H_7	H	2-Me	*−0.5306*	0.54	43.60	−30.9985
10	5-methyl-4-(2-methyl, 5-cloroindolyl)-2- methylguanidinothiazole	H	Me	2-Me, 5-Cl	0.5108	−1.17	35.87	−31.2836
11	5-methyl-4-(2-methyl, 5-cloroindolyl)-2- methylheptylguanidinothiazole	C_7H_7	Me	2-Me, 5-Cl	−0.0953	0.34	47.36	−39.1314
12	4-(4-methylindolyl)-2-methylheptylguanidinothiazole	C_7H_7	H	4-Me	−1.1939	0.66	43.60	−38.6181
13	4-(5-methylindolyl)-2-methylguanidinothiazole	H	H	5-Me	−0.5878	−0.85	32.10	−20.2655
14	4-(5-methylindolyl)-2-methylheptylguanidinothiazole	C_7H_7	H	5-Me	*0.0619*	0.66	43.60	−29.0775

Table 5.2. *(Continued)*

No.	Name	R1	R2	R3	$\lvert Y\rangle = A =$ Log(1/IC50)	$\lvert X_1\rangle =$ LogP	$\lvert X_2\rangle =$ POL[$Å^3$]	$\lvert X_3\rangle = E_{TOT}$ [kcal/mol]
	Compounds							
15	4-(6-methylindolyl)-2-methylheptylguanidinothiazole	C_7H_7	H	6-Me	*−0.4700*	0.66	43.60	−28.6606
16	4-(7-methylindolyl)-2-methylguanidinothiazole	H	H	7-Me	−2.0281	−0.85	32.10	−20.8977
17	4-(7-methylindolyl)-2-methylheptylguanidinothiazole	C_7H_7	H	7-Me	*0.0*	0.64	41.76	−27.5365
18	4-(5-cloroindolyl)-2-methylguanidinothiazole	H	H	5-Cl	−0.5878	−1.23	32.20	−21.3425
19	4-(5-cloroindolyl)-2-methylheptylguanidinothiazole	C_7H_7	H	5-Cl	0.3567	0.29	43.69	−29.6063
20	5-methyl-4-(5-cloroindolyl)-2-methyl-guanidinothiazole	H	Me	5-Cl	**0.5276: *max***	−1.20	34.03	−28.0026
21	5-methyl-4-(5-cloroindolyl)-2-methyl-heptylguanidinothiazole	C_7H_7	Me	5-Cl	0.3567	0.32	45.53	−37.9271
22	4-(5-bromoindolyl)-2-methylguanidinothiazole	H	H	5-Br	0.0408	−0.95	32.89	−21.1396
23	4-(5-bromoindolyl)-2-methylheptylguanidinothiazole	C_7H_7	H	5-Br	0.3567	0.56	44.39	−28.7985
24	4-(5-fluoroindolyl)-2-methylguanidinothiazole	H	H	5-F	−2.0015	−1.61	30.18	−20.5329
25	4-(5-fluoroindolyl)-2-methylheptylguanidinothiazole	C_7H_7	H	5-F	*−0.9933*	−0.09	41.67	−30.8076
26	4-(5-methylcetylindolyl)-2-methylguanidinothiazole	H	H	5-CO_2Me	*−1.9021*	−1.59	34.66	−27.4917
27	4-(5-carboximethylindolyl)-2- ethylbenzene-guanidinothiazole	C_7H_7	H	5-CO_2Me	−1.1314	−0.07	46.16	−37.7917
28	4-(5-amidoethylindolyl)-2- ethylbenzene-guanidinothiazole	C_7H_7	H	5-NHCOEt	**−3.1697: *min***	−0.75	48.71	−33.8852

The data is arranged in a vectorial manner, see the Table 5.1, so that to allow the direct implementation of SPECTRAL-SAR algorithm. However, as structural parameters the standard Hansch correlation (5.20) is here tested through evaluating the hydrophobic (*LogP*), electronic (*POL*arizability), and steric (total energy E_{TOT}) contribution on predicted PPI activity.

$$A = b_0 + b_1 \begin{pmatrix} hydrophobic \\ descriptor \end{pmatrix} + b_2 \begin{pmatrix} electronic \\ descriptor \end{pmatrix} + b_3 \begin{pmatrix} steric \\ descriptor \end{pmatrix}. \tag{5.20}$$

Before going into particular analysis and results, let us quote that for employed observed activity we use the logarithm of the inverse of the IC50 measured concentration since it allows better discrimination (in terms of sign) for the low and high activity (affinity for effector–receptor binding):

$$A = \log\left(1/IC_{50}\right) \begin{cases} < 0 \quad ... \quad IC_{50} > 1 \ ... \ low \ activity \\ > 0 \quad ... \quad 0 < IC_{50} < 1 \ ... \ high \ activity \end{cases}. \tag{5.21}$$

This rule further opens the way of deciding at once which molecule of Table 5.2 is the most active one; it follows that the molecule 20 (5-methyl-4-(5-cloroindolyl)-2-methylguanidinothiazole) displays the most positive record throughout the entire considered series. However, in molecular design we are searching for molecules that optimally act through controlled or predicted mechanisms. In this respect, we like to test whether the Hansch assumed mechanisms (seen as a combination of transduction by LogP, electrostatic interaction by POL, and fine tuning of covalent bindings by E_{TOT}) eventually leads with the same results or predict something different. Such endeavor can be fully accomplished by applying the SPECTRAL-SAR algorithm since its inclusion of the minimum spectral paths analysis. It produces the results presented in the Table 5.3 when the SPECTRAL-SAR recipe was respective engaged for various endpoint models for single-, double-, and all-structural effects in (5.20) included.

Table 5.3. QSAR equations through SPECTRAL-SAR multi-linear procedure for all possible correlation models considered from data of Table 5.2; $\|YMEASURED\| = 7.26044$.

Model	Variables	QSAR Equation	$\left\| Y \right\rangle^{PREDICTED} \right\|$	$r_{Algebraic}$	$r_{Statistic}$
Ia	$\left\|X_0\right\rangle$, $\left\|X_1\right\rangle$	$\left\|Y\right\rangle^{Ia} = -0.617953\left\|X_0\right\rangle + 0.644409\left\|X_1\right\rangle$	5.42199	0.746785	0.505266
Ib	$\left\|X_0\right\rangle$, $\left\|X_2\right\rangle$	$\left\|Y\right\rangle^{Ib} = -2.84242\left\|X_0\right\rangle + 0.0497704\left\|X_2\right\rangle$	4.9347	0.67967	0.306765
Ic	$\left\|X_0\right\rangle$, $\left\|X_3\right\rangle$	$\left\|Y\right\rangle^{Ic} = -2.26235\left\|X_0\right\rangle - 0.0478985\left\|X_3\right\rangle$	4.95654	0.682677	0.317809

Table 5.3. *(Continued)*

Model	Variables	QSAR Equation	$\left\| Y \right\rangle^{PREDICTED} \right\|$	$r_{Algebraic}$	$r_{Statistic}$
IIa	$\left\|X_0\right\rangle$, $\left\|X_1\right\rangle$, $\left\|X_2\right\rangle$	$\left\|Y\right\rangle^{IIa} = 1.2815\left\|X_0\right\rangle$ $+0.929486\left\|X_1\right\rangle - 0.0451665\left\|X_2\right\rangle$	5.50092	0.757657	0.531819
IIb	$\left\|X_0\right\rangle$, $\left\|X_1\right\rangle$, $\left\|X_3\right\rangle$	$\left\|X_0\right\rangle = -0.55429\left\|X_0\right\rangle$ $+0.655559\left\|X_1\right\rangle + 0.00204383\left\|X_3\right\rangle$	5.4223	0.746828	0.505372
IIc	$\left\|X_0\right\rangle$, $\left\|X_2\right\rangle$, $\left\|X_3\right\rangle$	$\left\|Y\right\rangle^{IIc} = -2.56541\left\|X_0\right\rangle$ $+0.0192239\left\|X_2\right\rangle - 0.0321241\left\|X_3\right\rangle$	4.96627	0.684018	0.322626
III	$\left\|X_0\right\rangle$, $\left\|X_1\right\rangle$, $\left\|X_2\right\rangle$, $\left\|X_3\right\rangle$	$\left\|Y\right\rangle^{III} = 2.33209\left\|X_0\right\rangle$ $+1.03407\left\|X_1\right\rangle - 0.120533\left\|X_2\right\rangle$ $-0.0680251\left\|X_3\right\rangle$	5.62022	0.774088	0.570269

Worth noted that while the statistical correlation factors (5.8) for computed models in Table 5.3 give quite modest values, indicating practically no correlation effects, the algebraic frame, specific to SPECTRAL-SAR quest, deliver systematically higher values, and better correlations accordingly, in each endpoint predicted case. This is a typical behavior within the vectorial analysis, previously consecrated on the enounced SPECTRAL-SAR correlation proposition.

Next, having the vectorial norms and algebraic (statistic) correlation factors one can unfold the spectral-path calculations, based on iterative application of the Eulerian formula (5.19). The hierarchy of alpha, beta, and gamma paths is established on the ground of the minimum path principle, both at global and local levels: first the minimum path is selected; if two paths are equal the minimum will be set along which the distance between the endpoint norms of the first two models of paths is minimum (this is based on heuristic principle of "first minimum movement" is the favorite one); if, by chance, also that distance is equal the chosen path will be that starts from the endpoint with the closest norm to the experimental observed one. In these conditions the results displayed in the Table 5.4. Note that, although no relevant for the SPECTRAL-SAR study the statistical correlation factors provide the same spectral path hierarchy as that algebraically based.

Table 5.4. Synopsis of paths connecting the SPECTRAL-SAR models of Table 5.3 in the norm-correlation spectral-space.

Path	Value	
	Algebraic	Statistic
Ia-IIa-III	0.200101	0.208619
Ia-IIb-III	α 0.200101	α 0.208615
Ia-IIc-III	1.12014	1.19022
Ib-IIa-III	γ 0.691988	γ 0.734646
Ib-IIb-III	0.691988	0.734596
Ib-IIc-III	0.691988	0.734781
Ic-IIa-III	0.669948	0.710281
Ic-IIb-III	0.669948	0.710398
Ic-IIc-III	β 0.669948	β 0.710129

However, this approach has to be further completed with a mechanism of identifying of the most active molecule through the selected spectral paths. This can be touched since employed two more stages in selecting the molecules along the spectral paths. Actually, in the first stage the residues of all predicted models' activities (computed upon the SPECTRAL-SAR equations of Table 5.3) against the observed activity of molecules of Table 5.2 are evaluated, and the Table 5.5 was produced. Nevertheless, this is due the huge role the residues acquire in SPECTRAL-SAR correlation, see the left side of (5.13) with the residues types (5.14) with the occasion of algebraic versus statistical previously discussion.

Table 5.5. Residual activities $A_i - Y_i^{Model}$ of the compounds of Table 5.2 for the SPECTRAL-SAR models of Table 5.3 ordered according with the alpha, beta, and gamma paths of Table 5.4.

No.	Models						
	α		β		γ		
	Ia	*IIb*	*Ic*	*IIc*	*Ib*	*IIa*	*III*
1	−0.934838	−0.956726	−0.652554	−0.695041	−0.861333	−1.18202	−0.865965
2	0.541102	0.543387	0.444569	0.409946	0.358506	0.673195	0.817532
3	−0.778626	−0.775505	−1.0748	−0.957714	−0.783313	−0.948854	−1.33679
4	−1.03763	−1.03878	−1.62443	−1.6404	−1.73147	−0.888172	−0.600457
5	−0.531331	−0.517643	−0.883578	−0.875296	−0.818025	−0.295776	−0.358289
6	0.764302	0.776127	0.444203	0.470903	0.549852	0.925302	0.800366
7	−0.6207	−0.604119	−0.469931	−0.496797	−0.428007	−0.373602	−0.51315
8	−0.947726	−0.955553	−0.987312	−0.954732	−0.952413	−1.11795	−1.14149
9	*XIV:−0.260628*	−0.266956	**XII:0.24697**	**VI:0.200851**	**III:0.141828**	−0.344759	−0.274524
10	1.88271	1.89603	1.27471	1.38169	1.56795	1.93693	1.58402
11	0.303554	0.316078	0.292716	0.302607	0.389992	0.446265	0.267553
12	−1.00126	−0.99335	−0.781297	−0.707222	−0.521472	−1.1196	−1.58024
13	0.577901	0.565135	0.703865	0.709513	0.656987	0.370612	0.449618
14	0.254543	**XI:0.242951**	0.931483	0.855061	0.734328	0.136203	0.324564
15	−0.277357	−0.289801	0.419552	0.336553	0.202428	−0.395697	*V:−0.178976*

Table 5.5. *(Continued)*

No.	α		β		γ		
	Ia	*IIb*	*Ic*	*IIc*	*Ib*	*IIa*	*III*
16	−0.862399	−0.873873	−0.766717	−0.751096	−0.783313	−1.06969	−1.03369
17	**VII:0.205531**	0.191012	0.943395	0.878036	0.764006	**I:0.00978637**	**IV:0.166392**
18	0.822776	0.816448	0.652278	0.672993	0.65201	0.728334	0.781355
19	0.787774	0.781388	1.20095	1.13114	1.02465	0.778978	0.976846
20	*1.91884*	*1.92579*	*1.44867*	*1.53926*	*1.67633*	*1.8985*	*1.63325*
21	0.768442	0.778728	0.802401	0.828474	0.933071	0.8342	0.601582
22	1.27094	1.26108	1.2906	1.29485	1.24627	1.12784	1.21739
23	0.613784	0.602736	1.23965	1.14364	0.989809	0.559633	0.836971
24	−0.346048	−0.349794	−0.722643	−0.675867	−0.661154	−0.423398	−0.427802
25	−0.31735	−0.317044	*VIII:−0.206586*	*IX:−0.218615*	*X:−0.224815*	−0.309054	−0.305402
26	0.259537	*XIII:−0.249283*	−0.956559	−0.886135	−0.784725	*II:−0.140242*	−0.282469
27	−0.468338	−0.453981	−0.679214	−0.667388	−0.586384	−0.262946	−0.398085
28	−2.06844	−2.05448	−2.5304	−2.62922	−2.7516	−1.55402	−1.16012

In the second stage, from the residues table there are selected the least positive and negative values for each endpoint along the optimized spectral paths (α, β, and γ) since that will target the best predicted-measured fit. Now, from the series of residue values is ordered from the smallest ahead and for each of it the respective observed activity is identified. The molecules with null and negative observed activities are considered unfavorable (or hard favorable) in binding. The search continues until the first molecule of the ordered set is found with the closest to the higher observed activity. In the present analysis, it results that the molecule 14 of Table 5.2, namely the 4-(5-methylindolyl)-2- methylheptylguanidinothiazole, fulfills such behavior. Note that if, by chance, there are two candidate molecules that furnish the same observed activity will be preferred that one belonging to the spectral pathway with superior degree (α>β>γ). If still, also by this constraint the molecules belongs to the same path (say they both display equal positive and negative residual values respecting the same observed activity) then the common skeleton resulting from their superposition will be considered the optimum-by-spectral-paths active molecule.

In fact, this kind of completing of SPECTRAL-SAR algorithm leads with the predicted 3D molecular structure that throughout a certain structural mechanism provides optimum attack on envisaged biological sites. It may be appropriately called as 3D-SPECTRAL-SAR method.

However, in the present case, there is clear that the resulted 3D-SPECTRAL-SAR molecule is significantly different by that simply picked up from Table 5.2 grounded only on the highest observed activity. In this regard, since the molecule of Table 5.2 with the smallest activity value, the molecule 28, 4-(5-amidoethylindolyl)-2-ethylbenzeneguanidinothiazole, still produces an observable activity, it means that the actual founded molecule, number 14 in Table 5.2, assure about 87.4%, of the maximum possible activity in the series.

Moreover, the molecule selected through the computational 3D-SPECTRAL-SAR model is expected to better satisfy the controlling and designing needs of the drugs along a specific pathway and interactions.

In other words, the molecular structure that produces an appreciable biological activity through the alpha spectral path chemical–biological interaction, that is through the dispersion (*LogP*), followed by steric-covalent adjustment of ligand-receptor (E_{TOT}), and in the final by electrostatic interaction (*POL*, here electrophile) attack—is that identified as molecule 14 out of series in Table 5.2.

This is not surprising at all since the representations in the Figure 5.3. At close inspection to the electrostatic contours for the molecules with the minimum, highest, and the 3D-SPECTRAL-SAR provided activities, respectively, there is clear that the last one is characterized by the largest area of electrophile attack the entirely covers the basic benzothiazolyl ring of the three analyzed compounds. Through this feature it seems that the 4-(5-methylindolyl)-2- methylheptylguanidinothiazole molecule better responds to the inner mechanisms of protonation upon entering the lower pH gradient that is present near the secretory canaliculus. The molecules with minimum and maximum observed activities indeed show somewhat complementary electrophile/nucleophile areas in Figure 5.3, (a) and (b), respectively, being perhaps involved in many indirect or complex signaling pathways.

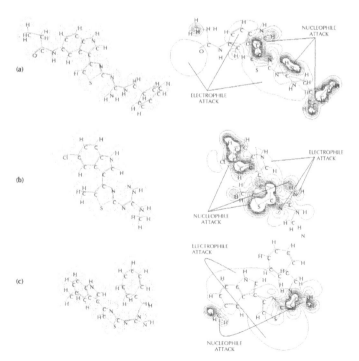

Figure 5.3. The electronic basins (in left sides) and the associated electrostatic potential contours (in right sides) for (a) molecule 28 of Table 5.2, emphasizing minimum action and maximum misfit; (b) molecule 20 of Table 5.2, with maximum observed action and absolute minimum misfit; and (c) molecule 14 of Table 5.2, displaying optimized maximum activity by predicted mechanism action.

Further studies may concern upon finding the appropriate structural factors that be-ing involved in the 3D-SPECTRAL-SAR correlation and path analysis to predict the molecular mechanism recovering the maximum observed activity at the best.

CONCLUSION

In quantitative structure-activity relationship studies, among the search for the best correlation of predicted with the observed data, stands also the need of identifying the molecular structure, as the 3D configuration, that fits at the best with the envis-aged receptor throughout a given (thus susceptible of being controlled) mechanism of action. This way the molecular design for a specific pathway, chemical–biological interaction, and target can be assured for the new synthesized chemicals and drugs.

However, unfortunately, until now there was not approached the 3D predicted structure from the correlation itself but only as a prerequisite for further correlation through validations and tests. In this context, this work opens the way from which the optimized 3D molecular for ligand-receptor bet fit to be extracted as a result of QSAR performed correlations on a given series of analogs.

The SPECTRAL-SAR algorithm was here proved that supports extension such that to deliver such 3D structural information. It starts with the consecrated vectorial analysis [10-13] while providing the main least spectral paths connecting the con-sidered endpoint computed models for a given model of correlation analysis, here of Hansch type. Then, for each proposed model the residues of the predicted activities against the measured one, foe each molecule of a tested series, is employed to build a series of minimum-to-maximum series of compounds that candidate for the optimized 3D structure. The final cut is decided for that molecule in the residues ordered series that displays the closest (positive) activity to the maximum observed in the series. This so called 3D-SPECTRAL-SAR method allows for further identification of molecular structure that is intrinsically controlled by the envisaged structural factors (here LogP, POL, and E_{TOT}) and of the associated pathways toward the bio-chemical, receptor, or substrate target. However, this by-SPECTRAL-SAR predicted optimized molecular structure may often be different by the molecule in a given series displaying the maxi-mum observed activity.

This was the case also for the actual study of the inhibitory activity of 4-indolyl, 2-guanidinothiazole derivatives upon the H^+K^+-ATPase proton pump in parietal cells of the stomach. Actually, there was found that the molecule selected by the 3D-SPEC-TRAL-SAR method shows a larger area of electrophilic interaction that directly links with the protonation role that optimal H^+K^+-ATPase inhibitory has to have, whereas the molecule with the highest observed activity in the considered series poses a rather equilibrium among the electrophilic and nucleophilic active sites suggesting that it is involved in a more complex physico-chemical processes in the cell until the activity signaling is produced.

Nevertheless, at this point the temporal issue of the chemical–biological interac-tion may be assessed in order to decide whether the 3D molecular by-SPECTRAL-SAR algorithm is associated also with the shortest temporal range of action or otherwise. Such studies may be unfolded appealing also the recent logistical version of the enzyme-substrate kinetics [13, 17–19] and are currently in progress.

KEYWORDS

- **Hydrophobicity**
- **Polarizability**
- **Proton-pump inhibitors**
- **Spectral paths**
- **SPECTRAL-SAR algorithm**

PERMISSIONS

This Chapter was previously published in International Journal of Chemical Modeling, 1(1) (2008) 45-62 as Turning SPECTRAL-SAR into 3D-QSAR Analysis. Application on H+K+-ATPase Inhibitory Activity.

PART II
SPECTRAL-SAR ASSESSMENT ON IONIC LIQUIDS' TOXICITY

Chapter 6

SPECTRAL-SAR Ecotoxicology of Ionic Liquids: The *Vibrio fischeri* Case

Ana-Maria Putz, Mihai V. Putz, and Vasile Ostafe

INTRODUCTION

Within the recently launched the spectral-structure activity relationship (SPECTRAL-SAR) analysis, the vectorial anionic-cationic model of a generic ionic liquid (IL) is proposed, along with the associated algebraic correlation factor in terms of the measured and predicted activity norms. The reliability of the present scheme is tested by assessing the Hansch factors, that is, lipophylicity, polarizability, and total energy, to predict the ecotoxicity endpoints of wide types of ILs with ammonium, pyridinium, phosphonium, choline, and imidazolium cations on the aquatic bacteria *Vibrio fischeri*. The results, while confirming the cationic dominant influence when only lipophylicity is considered, demonstrate that the anionic effect dominates all other more specific interactions. It was also proved that the SPECTRAL-SAR vectorial model predicts considerably higher activity for the ILs than for its anionic and cationic subsystems separately, in all considered cases. Moreover, through applying the least norm-correlation path principle, the complete toxicological hierarchies are presented, unfolding the ecological rules of combined cationic and anionic influences in IL toxicity.

Since their emergence a decade ago, ILs have had a constantly growing influence on organic, bio- and green chemistry, due to the unique physico-chemical properties manifested by their typical salt structure: a heterocyclic nitrogen-containing organic cation (in general) and an inorganic or organic anion [1], with melting points below 100°C and no vapor pressure [2]. The latter property leads to the practical replacement of conventional volatile organic compounds (VOCs) from the point of view of atmospheric emissions, though they do present the serious drawback that a small amount of IL could enter the environment through groundwater [3]. This risk makes it necessary to perform further ecotoxicological studies of IL on various species, in order to improve the "design rules" for synthesized IL with minimal toxicity to environment integrated organisms.

ILs display variable stability in terms of moisture and solubility in water, polar and nonpolar organic solvents [4]. Various values of IL hydrophobicity and polarity may be tailored [3] with the help of nucleoside chemistry [5] according to the main principles of green chemistry [6, 7]: the new chemicals must be designed to preserve

Lacrămă A.M., Putz M.V., Ostafe V. "A Spectral-SAR Model for the Anionic-Cationic Interaction in Ionic Liquids: Application to *Vibrio fischeri* Ecotoxicity", *International Journal of Molecular Sciences*, 8 (2007) 842-863.

effectiveness of function while reducing toxicity, and not persisting in the environment at the end of their usage, but breaking down into inoffensive degradation products.

Most of the ILs with imidazolium, phosphonium, pyridinium, and ammonium that were tested were resistant to ready biodegradation [8]. Their toxicity to microorganisms can limit biodegradation, while their toxicity to humans and others organisms is obviously significant. The examination regarding the biodegradation of surfactant compounds focuses on a close resemblance between many quaternary ammonium compounds as well as surfactants based around an imidazolium core. The factors that improved the biodegradation of surfactants have successfully been applied to ILs. For instance, bis(trifluoromethylsulfonyl)imide (TFMSi) and PF6 ILs containing an ester in the side chain exhibit the same hydrophobic character as 1-n-butyl-3-methylimidazolium bis(trifluoromethylsulfonyl)imide ([BMIM][TFMSi]) and 1-n-butyl-3-methylimidazolium hexafluorophosphate ([BMIM][PF6]). The enzymatic hydrolysis step, which initiates a pathway to further breakdown products, improves the biodegradation. Therefore, compound stability and toxicity are the factors biodegradability depends on. The effect of the counter-ion was not noticeable in biodegradability even though modifications of the anion led to changes in physical and chemical properties. Still, the introduction of an organic anion clearly improves the extent of ultimate biodegradation [9].

From the point of view of reactivity, IL generally do not coordinate to metal complexes, enzymes and different organic substrates [10]; however, they are usually the major component of the mixtures having pre-organized structures with the aid of many hydrogen bonds (structural directionality) in contrast to classical salts in which the compounds are mostly formed with the aid of ionic bonds (charge-ordering structures) [11, 12].

On the other way, the recycling ability of IL, especially dialkyl-imidazolium based ILs (the most studied until now), is based on their lack of solubility in some key organic solvents (e.g., diethyl ether) and in water—for the special case of 1-n-butyl-3-methylimidazolium hexafluorophosphate ([BMIM][PF6]). Water/soluble ILs are more difficult to recycle, since their water immiscible complements, the secondary products, cannot be easily removed [13].

The detailed examination of relative energies and structural interactions (like ion position, H-bonding, and anion conformational variability) in gas-phase ion-pairs has emphasized the way these quantities can be used to build up a picture of the local structure and interactions occurring in ILs. For instance, while 1-n-butyl-3-methylimidazolium chloride ([BMIM][Cl]) forms a highly connected liquid with relatively strong interactions—at one extreme, and 1-n-butyl-3-methylimidazolium bis(trifluoromethylsulfonyl)imide ([BMIM][TFMSi]) forms a low connectivity network of weakly linked ions—at the other extreme, [BMIM][BF4] lying between these two extremes forms a weak but more regular network. Melting points and viscosity are partly dependent on local interactions between an ion and other molecules in the first solvation shell. For imidazolium based cations, nine sites of interactions are preferred by the anion [14]. Hunt et al. proved that the hydrogen bond is primarily ionic with a moderate covalent character [15]. The fact that the Coulombic attraction is the dominant stabilization force was demonstrated by the analysis of the charge distribution, molecular orbitals, and electron density of the dimer complex 1-n-butyl-3-methylimidazolium chloride ([BMIM][Cl]): the interactions governing the top conformers are very different from those in which the Cl anion remains in plane (were Cl anion interacts with the p manifold of orbitals); the ion-pair LUMO is the cation anti-bonding LUMO and thus electron acceptance is not favorable [15]. The effect of the chloride anion on rotation of the butyl chain is investigated and found to lower some rotational barriers while enhancing others [16]. Since the Coulombic forces are significant in an IL, charge distribution or point charges on the constituent ions are likely to be more significant than in liquids made up of neutral charge molecules [17] and deserve a more detailed study.

In this respect, the costs of all approaches for sustainable product design can be reduced using structure-activity relationship (SAR) and quantitative structure-activity relationship (QSAR) methods [18]. It has already been proved that the anti-microbial activity of quaternary ammonium chlorides is lipophilicity-dependent [19]. While the 1-octanol-water partition coefficient could be seen only as the first approximation for compound lipophylicity, bioaccumulation, and toxicity in fish, as well as sorption to soil and sediments assumes that lipophylicity is the main factor of anti-microbial activity [20]. Nevertheless, aiming at a deeper understanding of the specific mechanistic description of IL ecotoxicity, it is worth considering that the IL properties are more comprehensively quantified through lipophylicity, polarizability, and total energy as a unitarily complex of factors in developing appropriate SAR studies.

However, the main problem in assessing the viable QSAR studies to predict IL toxicities concerns the anionic-cationic interaction superimposed on the anionic and cationic subsystems containing ILs. There are basically two complementary ways of attaining this goal. One may address the search of special rules for assessing the anionic-cationic structural separately from the individual anionic and cationic ones, and then generating the QSAR models. Yet, because the cationic and anionic effects on liquid toxicity are merely separately studied at the moment, the appropriate strategy would be to firstly derive the anionic and cationic QSARs and only then to move on to a QSAR of the IL viewed as an anionic-cationic interaction. The current paper shows, for the first time, how the latter procedure may be implemented by means of the vectorial approach of the SPECTRAL-SAR analysis. The illustration of the SPECTRAL-SAR-IL model presented is performed by studying the aquatic bacteria *Vibrio fischeri* toxicity against a list of 22 ILs, appropriately chosen so that they should contain a wide variety of heads, side chains, and anions. This way, the present methodology and results may be extended over a wide range of organisms towards designing specific ecotoxicological IL batteries.

THE SPECTRAL-SAR IONIC LIQUID (SPECTRAL-SAR-IL) MODEL

Although IL structure is currently defined through three types of substructures, the head group (the positively charged moiety) with the side chain, R1, R2, and so forth, which are substituents on heads group, and the anion [7], the present approach views IL as consisting mainly of its cationic (head and side chain) and anionic subsystems in mutual interaction. This way, the two sub-systems containing IL can be modeled through their vectorial activities that sum up into the predicted activity of the IL as a whole. This picture is further sustained by the possibility of employment of the SPECTRAL-SAR method to the present purposes.

SPECTRAL-SAR Concepts

Without going into details [18], if one has to solve the correlation between a set of biological activities of N-compounds with the set of M-structural properties of each of them,

$$|Y\rangle = b_0|X_0\rangle + b_1|X_1\rangle + ... + b_k|X_k\rangle + ... + b_M|X_M\rangle , \qquad (6.1)$$

a spectral algebraic algorithm can be applied based on the vectorial view of the descriptors and measured activity combined with the Gram–Schmidt orthogonalization recipe and on the scalar product rule

$$\langle \Psi_l | \Psi_k \rangle = \sum_{i=1}^{N} \psi_{il} \psi_{ik} = \langle \Psi_k | \Psi_l \rangle \qquad (6.2)$$

giving out a real number from any two arbitrary N-dimensional vectors

$$|\Psi_l\rangle = |\psi_{1l} \quad \psi_{2l} \quad ... \quad \psi_{Nl}\rangle, |\Psi_k\rangle = |\psi_{1k} \quad \psi_{2k} \quad ... \quad \psi_{Nk}\rangle . \qquad (6.3)$$

Let's also note that when considering the N-biological activities as well as their respective predictor variables grouped into vectors, as displayed in Table 6.1, the unity vector $|X_0\rangle = |1 \quad 1 \quad ... \quad 1\rangle$ was added to account for the free correlation term in (6.1).

Table 6.1. The vectorial descriptors in a SPECTRAL-SAR analysis.

Activity		Structural Predictor Variables				
$|Y\rangle$	$|X_0\rangle$	$|X_1\rangle$	...	$|X_k\rangle$	...	$|X_M\rangle$
y_1	1	x_{11}	...	x_{1k}	...	x_{1M}
y_2	1	x_{21}	...	x_{2k}	...	x_{2M}
⋮	⋮	⋮	⋮	⋮	⋮	⋮
y_N	1	x_{N1}	...	x_{Nk}	...	x_{NM}

Actually, the SPECTRAL-SAR equation is derived from the determinant [18]:

$$\begin{vmatrix} |Y\rangle & \omega_0 & \omega_1 & \cdots & \omega_k & \cdots & \omega_M \\ |X_0\rangle & 1 & 0 & \cdots & 0 & \cdots & 0 \\ |X_1\rangle & r_0^1 & 1 & \cdots & 0 & \cdots & 0 \\ \vdots & \vdots & \vdots & \vdots & & \vdots & \\ |X_k\rangle & r_0^k & r_1^k & \cdots & 1 & \cdots & 0 \\ \vdots & \vdots & \vdots & \vdots & & \vdots & \\ |X_M\rangle & r_0^M & r_1^M & \cdots & r_k^M & \cdots & 1 \end{vmatrix} = 0, \qquad (6.4)$$

once it is expanded along its first column, and where

$$r_i^k = \frac{\langle X_k | \Omega_i \rangle}{\langle \Omega_i | \Omega_i \rangle}, \omega_k = \frac{\langle \Omega_k | Y \rangle}{\langle \Omega_k | \Omega_k \rangle}, |\Omega_k \rangle = |X_k \rangle - \sum_{i=0}^{k-1} r_i^k |\Omega_i \rangle, k = \overline{0, M}. \qquad (6.5)$$

With these ingredients the QSAR equation (6.1) becomes operational, with all coefficients effectively worked out. However, besides the effectiveness of the SPEC-TRAL-SAR methodology in reproducing the old-fashioned multi-linear QSAR analysis, one of its advances concerns the possibility of introducing norms associated with either predicted (computed) or experimental (measured) activities,

$$\| Y \rangle \| = \sqrt{\langle Y | Y \rangle} = \sqrt{\sum_{i=1}^{N} y_i^2} . \qquad (6.6)$$

Thus, gaining the possibility of the unique assignment of a number to a specific type of correlation, that is performing a sort of final quantification of the models. Nevertheless, the activity norm given in equation (6.6) opens the possibility of replacing the classical statistical correlation factor

$$r_{S-SAR}^{STATISTIC} = \sqrt{1 - \frac{\sum_{i=1}^{N} (y_{iEXP} - y_i)^2}{\sum_{i=1}^{N} \left(y_{iEXP} - \frac{1}{N} \sum_{i=1}^{N} y_{iEXP} \right)^2}} \qquad (6.7)$$

with a new definition, introducing the so called *algebraic SPECTRAL-SAR correlation factor* as the ratio of the spectral norm of the predicted activity versus that of the measured one:

$$r_{S-SAR}^{ALGEBRAIC} = \sqrt{\frac{\sum_{i=1}^{N} y_i^2}{\sum_{i=1}^{N} y_{iEXP}^2}} \qquad (6.8)$$

In fact this new correlation factor definition compares the vectorial lengths of the predicted activity against the measured one, thus being an indicator of the extent to which certain computed property or activity approaches the "dimension" of the observed quantity. However, it was also shown that practically the algebraic correlation factor (6.8) furnishes higher values than its statistical counterpart definition (6.7), in a systematical manner [18], thus making it the ideal tool for the present attempt to consider the IL activity taken from its cationic and anionic sub-system ones.

Moreover, with the help of both spectral norms and correlation factors we can introduce the so called *the least path principle* in terms of paths between the tested models (or endpoints) [18]:

$$\delta[A,B] = 0; \quad A,B : ENDPOINTS \tag{6.9}$$

with

$$[A,B] = \sqrt{\left(\left\|\left|Y^B\right\rangle\right\| - \left\|\left|Y^A\right\rangle\right\|\right)^2 + \left(r_B^{\substack{STATISTIC/ \\ ALGEBRAIC}} - r_A^{\substack{STATISTIC/ \\ ALGEBRAIC}}\right)^2} \tag{6.10}$$

thus, providing a practical tool for deciding the dominant hierarchies along the possible ones with the important consequence of picturing the mechanistic and almost temporal evolution of structural causes that trigger the observed effects. This methodology was successfully applied in ecotoxicology [18] and for designing the behavior of the species interactions within a test battery [21], promising to furnish the frame also in ILs analysis.

SPECTRAL-SAR for Ionic Liquids

Basically, since we may consider the IL as composed by the anionic and cationic subsystems, we can employ the spectral vectorial space of anionic and cationic activities to form the IL one throughout the vectorial resultant

$$\left|Y_{AC}\right\rangle = \left|Y_C\right\rangle + \left|Y_A\right\rangle \tag{6.11}$$

as displayed also in Figure 6.1.

The vectorial summation in (6.11) may also be seen as the interference of the anionic and cationic activities in the space of IL so that their mutual interaction is included. Actually, searching for the predicted norm of the vectorial IL activity

$$\left\|\left|Y_{AC}\right\rangle\right\| = \sqrt{\langle Y_{AC} | Y_{AC}\rangle} = \sqrt{\sum_{i=1}^{N} y_{iAC}^2} \tag{6.12}$$

one can notice the interference effects between the anionic and cationic systems

$$\left\|\left|Y_{AC}\right\rangle\right\| = \sqrt{\left\|\left|Y_C\right\rangle\right\|^2 + \left\|\left|Y_A\right\rangle\right\|^2 + 2\left\|\left|Y_C\right\rangle\right\|\left\|\left|Y_A\right\rangle\right\|\cos\theta_{AC}} \tag{6.13}$$

through the correlation angle between the associated vectors, $\cos\theta_{AC}$. Nevertheless, its practical definition can be achieved, since the anionic-cationic norm (6.13) is rewritten employing the vectorial relation (6.11) to the scalar product between the anionic and cationic vectors.

$$\left\|\left|Y_{AC}\right\rangle\right\| = \sqrt{\left\|\left|Y_C\right\rangle\right\|^2 + \left\|\left|Y_A\right\rangle\right\|^2 + 2\langle Y_C | Y_A\rangle} \tag{6.14}$$

Now, from the two equivalent expressions (6.13) and (6.14), the anionic-cationic correlation angle results in a numerical form from the vectorial components of the anionic and cationic activities.

$$\cos \theta_{AC} = \frac{\sum_{i=1}^{N} y_{iC} y_{iA}}{\sqrt{\sum_{i=1}^{N} y_{iC}^2 \sum_{i=1}^{N} y_{iA}^2}} \qquad (6.15)$$

Basically, the degree of anionic-cationic activity interaction in IL is fixed by the value of the angle (6.15) for each envisaged QSAR model or endpoint.

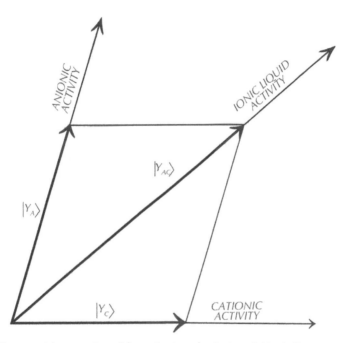

Figure 6.1. The vectorial summation of the cationic and anionic activities in IL.

Finally, aiming to obtain the IL working QSAR equation from the cationic and anionic counterparts one simply needs to expand the norm and the scalar product in (6.14) as the square sum of components, according to the norm and scalar product definitions (6.12) and (6.2), respectively, with the intermediate result:

$$\sqrt{\sum_{i=1}^{N} y_{iAC}^2} = \sqrt{\sum_{i=1}^{N} y_{iC}^2 + \sum_{i=1}^{N} y_{iA}^2 + 2\sum_{i=1}^{N} y_{iC} y_{iA}} \qquad (6.16)$$

Hence, by the direct identification of right and left side terms the quested equation arises

$$y_{iAC} = \sqrt{y_{iC}^2 + y_{iA}^2 + 2y_{iC}y_{iA}} = y_{iC} + y_{iA} \qquad (6.17)$$

With these all the required concepts and analytical tools are given for that any IL anionic-cationic interaction and resulted activity analysis on certain species or organism be performed, while concrete example will in next be exposed.

APPLICATION TO ECOTOXICOLOGY

The Working System

Usually, the classical assessment of new industrial chemicals is *inflexible,* being directed by regulations and standardized procedures, while biological test systems are very expensive and the costs increase with the number of new chemicals released. So, a *flexible test strategy* is needed in correlating the chemical compounds with the systems they act upon. Although, qualitative (eco) toxicological algorithms for selection of test systems have been proposed in terms of identification of individual effects of different head-groups (with identical side chains and anions) and different anions (with identical cations) [7], a quantitative *theoretical prediction algorithm* for tested compounds in biological systems is still need to be proposed due the lack of readily accessible ecotoxicological data [8].

The principles of green chemistry say that one has to consider the whole process (life cycle analysis) rather than individual components of reaction (single issue sustainability), based on the fact that the acute toxicity measurements do not provide a complete characterization of the full impact of a substance release into environment, but are only part of the environmental impact assessment [6, 8].

In this context, ILs with cations like pyridinium, imidazolium, and pyrrolidinium have already been nominated in the United States National Toxicology Program (NTP) for toxicological testing based upon their potential of new solvents but also due they ability to enter in aquatic system [22]; if an accidentally discharge of ILs into water occur, many of them being water soluble, they may be an environmental risk to aquatic plants and animals [23].

For this reason, the current application will develop the SPECTRAL-SAR complete algorithm up to the mechanistic prediction of the correlated structural causes and ecotoxicological effects for the ILs of Figure 6.2, containing ammonium, pyridinium, phosphonium, choline, and imidazolium cations, on aquatic bacteria *Vibrio fischeri* [24]. Whereas, the *Vibrio fisheri* species was previously found to be one of the most resistant species to ordinary phenol compounds toxicity [21], the present ILs show quite a wide structural variety to furnish useful information of their environmental action. The cationic and anionic structural properties, the lipophylicity, the electronic polarizability, and the total energy, where computed with the HyperChem computational environment [25] and displayed in Table 6.2 together with the reported measured activity of their containing ILs, respectively.

The data in Table 6.2 are suitable arranged so that the SPECTRAL-SAR analysis is performed successively at the cationic and anionic level and then at the IL levels based on equations (6.4)–(6.10) and (6.15)–(6.17) in a Hansch type expansion:

$$A = f_{Linear}\left(\log P, POL, E_{TOT}\right),\qquad(6.18)$$

when accounting for different combinations between the lipophylicity (hydrophobic character), polarizability (electronic character) and total energy (steric character) factors, respectively. The spectral hierarchy of the predicted activities with respect to different concerned endpoints would lead to the mechanistic scheme according with the cationic and anionic sides as well as the overall ILs influences the environmental (*Vibrio fischeri*) toxicity.

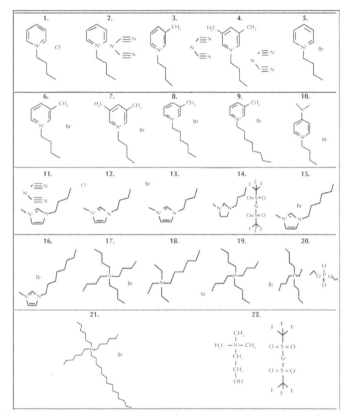

Figure 6.2. The ionic liquids acting on *Vibrio fischeri* species studied in this work.

Table 6.2. The series of the ILs of Figure 6.2 of those toxic activities $A = Log(EC_{50})$ on *Vibrio fischeri* were considered [24], with the marked values taken from [4], along structural parameters *LogP, POL* (E^3), and E_{TOT} (kcal/mol) as accounting for the hydrophobicity, electronic (polarizability), and steric (total energy at optimized 3D geometry) effects, computed with the help of HyperChem program [25], for each cation and anion containing ionic liquid, respectively.

No.	NAME	A_{exp}	Log P		Polarizability		TOTAL ENERGY	
		$\lvert Y_{EXP} \rangle$	CAT.	AN.	CAT.	AN.	CAT.	AN.
			$\lvert X_{1C} \rangle$	$\lvert X_{1A} \rangle$	$\lvert X_{2A} \rangle$	$\lvert X_{2A} \rangle$	$\lvert X_{3C} \rangle$	$\lvert X_{3A} \rangle$
1.	1-n-butylpyridinium chloride	0.41*	2.85	0.63	17.51	2.32	–250008.14	–285190.78
2.	1-n-butylpyridinium dicyano-amide	0.31*	2.85	0.43	17.51	5.51	–250008.14	–147935.98

Table 6.2. *(Continued)*

No.	NAME	A_{exp} $\lvert Y_{EXP}\rangle$	Log P CAT. $\lvert X_{1C}\rangle$	AN. $\lvert X_{1A}\rangle$	Polarizability CAT. $\lvert X_{2A}\rangle$	AN. $\lvert X_{2A}\rangle$	TOTAL ENERGY CAT. $\lvert X_{3C}\rangle$	AN. $\lvert X_{3A}\rangle$
3.	1-n-butyl-3-methylpyridinium dicyanoamide	−0.34*	3.32	0.43	19.35	5.51	−274222.62	−147935.98
4.	1-n-butyl-3,5-dimethylpyridinium dicyanoamide	−0.62*	3.78	0.43	21.18	5.51	−298437.03	−147935.98
5.	1-n-butylpyridinium bromide	0.40*	2.85	0.94	17.51	3.01	−250008.14	−1596918.25
6.	1-n-butyl-3-methylpyridinium bromide	−0.25*	3.32	0.94	19.35	3.01	−274222.62	−1596918.25
7.	1-n-butyl-3,5-dimethylpyridinium bromide	−0.31*	3.78	0.94	21.18	3.01	−298437.03	−1596918.25
8.	1-n-hexyl-3-methylpyridinium bromide	−0.94*	4.11	0.94	23.02	3.01	−322641.81	−1596918.25
9.	1-n-octyl-3-methylpyridinium bromide	−2.21*	4.90	0.94	26.69	3.01	−371060.81	−1596918.25
10.	1-n-butyl-4-dimethylamino-pyridinium bromide	−0.68	3.11	0.94	22.53	3.01	−332525.97	−1596918.25
11.	1-n-butyl-3-methylimidazolium dicyanoamide	0.67*	0.68	0.43	17.22	5.51	−260646.64	−147935.98
12.	1-n-butyl-3-methylimidazolium chloride	0.71*	0.68	0.63	17.22	2.32	−260646.64	−285190.78
13.	1-n-butyl-3-methylimidazolium bromide	1.01*	0.68	0.94	17.22	3.01	−260646.64	−1596918.25
14.	1-n-butyl-3-methylimidazolium bis(trifluoromethanesulfonyl)imide	0.39	0.68	3.05	17.22	7.20	−260646.64	−1128283.62
15.	1-n-hexyl-3-methylimidazolium bromide	−1.58*	1.47	0.94	20.89	3.01	−309065.84	−1596918.25
16.	1-n-octyl-3-methylimidazolium bromide	−2.37*	2.26	0.94	24.56	3.01	−357484.59	−1596918.25

Table 6.2. *(Continued)*

No.	NAME	A_{exp}	Log P		Polarizability		TOTAL ENERGY	
		$\lvert Y_{EXP}\rangle$	CAT.	AN.	CAT.	AN.	CAT.	AN.
			$\lvert X_{1C}\rangle$	$\lvert X_{1A}\rangle$	$\lvert X_{2A}\rangle$	$\lvert X_{2A}\rangle$	$\lvert X_{3C}\rangle$	$\lvert X_{3A}\rangle$
17.	Tetrabutylammonium bromide	0.27	4.51	0.94	30.91	3.01	−422421.97	−1596918.25
18.	hexyltriethylammonium bromide	−0.54	2.71	0.94	23.57	3.01	−325587.25	−1596918.25
19.	tetrabutylphosphonium bromide	−0.29	2.89	0.94	30.91	3.01	−600149.62	−1596918.25
20.	tributylethylphosphonium diethylphosphate	0.07	2.02	2.63	27.24	10.53	−551729.87	−494172.37
21.	Trihexyl(tetradecyl)phosphonium bromide	0.41	9.23	0.94	60.27	3.01	−987499.25	−1596918.25
22.	Choline bis(trifluoromethanesulfonyl)imide	1.15	−0.76	3.05	11.36	7.20	−202450.36	−1128283.62

Results and Discussion

As earlier mentioned, the first step in our analysis consists in deriving the cationic and anionic QSARs that link the structural lipophilic-electronic-steric parameters of the ILs of Figure 6.2 and Table 6.2 with the observed activities of the whole containing ILs upon the *Vibrio fischeri* species. The results are presented in Tables 6.3 and 6.4 for the cationic and anionic subsystems for all main combinations, that is, generating the uni-modes *Ia*-to-*Ic* when only one structural parameter is correlated, the two-modes *IIa*-to-*IIc* when two combined structural factors are taken into account and for the three-mode correlation *III* with all structural factors involved, respectively. For each such mode of action, the associated endpoint norm, the statistic and algebraic correlation factors were reported, computed with the equations (6.6), (6.7), and (6.8), respectively.

Table 6.3. Spectral structure activity relationships (SPECTRAL-SAR) of the ionic liquids of Figure 6.2 against *Vibrio fischeri* toxicity, and the associated computed spectral norms with $\lVert \lvert Y_{EXP}\rangle \rVert = 4.41537$, statistic and algebraic correlation factors, computed upon the relations (6.6), (6.7), and (6.8), throughout the possible correlation models considered from the *cationic data* in Table 6.2, respectively.

Mode	Vectors	Cationic SPECTRAL-SAR	$\lVert Y_A\rangle^{Mode}\rVert$	$r^{STATISTIC}_{S-SAR}$	$r^{ALGEBRAIC}_{S-SAR}$
Ia	$\lvert X_0\rangle,$ $\lvert X_{1A}\rangle$	$\lvert Y_A\rangle^{Ia} = 0.152926\lvert X_0\rangle$ $-0.124263\lvert X_{1C}\rangle$	1.47807	0.267342	0.334755

Table 6.3. *(Continued)*

Ib	$\lvert X_0 \rangle$, $\lvert X_{2C} \rangle$	$\lvert Y_C \rangle^{Ib} = 0.0998011 \lvert X_0 \rangle$ $-0.0129369 \lvert X_{2C} \rangle$	1.08531	0.132169	0.245803
Ic	$\lvert X_0 \rangle$, $\lvert X_{3C} \rangle$	$\lvert Y_C \rangle^{Ic} = -0.106319 \lvert X_0 \rangle$ $+2.57881 \cdot 10^{-7} \lvert X_{3C} \rangle$	0.9452	0.0469985	0.21407
IIa	$\lvert X_0 \rangle$, $\lvert X_{1C} \rangle, \lvert X_{2C} \rangle$	$\lvert Y_C \rangle^{IIa} = -0.195021 \lvert X_0 \rangle$ $-0.241666 \lvert X_{1C} \rangle +0.0295874 \lvert X_{2C} \rangle$	1.64279	0.314715	0.372062
IIb	$\lvert X_0 \rangle$, $\lvert X_{1C} \rangle, \lvert X_{3C} \rangle$	$\lvert Y_C \rangle^{IIb} = -0.139788 \lvert X_0 \rangle$ $-0.222688 \lvert X_{1C} \rangle -1.62349 \cdot 10^{-6} \lvert X_{3C} \rangle$	1.72651	0.3379	0.391023
IIc	$\lvert X_0 \rangle$, $\lvert X_{2C} \rangle, \lvert X_{3C} \rangle$	$\lvert Y_C \rangle^{IIc} = 0.384457 \lvert X_0 \rangle$ $-0.124625 \lvert X_{2C} \rangle -6.48599 \cdot 10^{-6} \lvert X_{3C} \rangle$	1.71867	0.335748	0.389246
III	$\lvert X_0 \rangle$, $\lvert X_{2C} \rangle, \lvert X_{2C} \rangle$, $\lvert X_{3C} \rangle$	$\lvert Y_C \rangle^{III} = 0.141278 \lvert X_0 \rangle$ $-0.127251 \lvert X_{1C} \rangle -0.0677301 \lvert X_{2C} \rangle$ $-4.48229 \cdot 10^{-6} \lvert X_{3C} \rangle$	1.79053	0.355322	0.405522

Table 6.4. Spectral structure activity relationships (SPECTRAL-SAR) of the ionic liquids of Figure 6.2 against *Vibrio fischeri* toxicity, and the associated computed spectral norms with $\lVert \lvert Y_{EXP} \rangle \rVert = 4.41537$, statistic and algebraic correlation factors, computed upon the relations (6.6), (6.7), and (6.8), through all possible correlation models considered from the *anionic data* in Table 6.2, respectively.

Mode	Vectors	Anionic SPECTRAL-SAR	$\lVert Y_A \rangle^{Mode} \rVert$	$r^{STATISTIC}_{S-SAR}$	$r^{ALGEBRAIC}_{S-SAR}$
Ia	$\lvert X_0 \rangle$, $\lvert X_{1A} \rangle$	$\lvert Y_A \rangle^{Ia} = -0.514106 \lvert X_0 \rangle$ $+0.291698 \lvert X_{1A} \rangle$	1.38453	0.238974	0.313569

Table 6.4. *(Continued)*

Mode	Vectors	Anionic SPECTRAL-SAR	$\left\|\left. Y_A \right\rangle^{Mode}\right\|$	$r_{S-SAR}^{STATISTIC}$	$r_{S-SAR}^{ALGEBRAIC}$
Ib	$\left\|X_0\right\rangle,$ $\left\|X_{2A}\right\rangle$	$\left\|Y_A\right\rangle^{Ib} = -0.703086\left\|X_0\right\rangle$ $+ 0.122745\left\|X_{2A}\right\rangle$	1.48745	0.270118	0.33688
Ic	$\left\|X_0\right\rangle,$ $\left\|X_{3A}\right\rangle$	$\left\|Y_A\right\rangle^{Ic} = 0.376373\left\|X_{3A}\right\rangle$ $+ 5.11098\cdot10^{-7}\left\|X_{3A}\right\rangle$	1.75453	0.345553	0.397368
IIa	$\left\|X_0\right\rangle,$ $\left\|X_{1A}\right\rangle,\left\|X_{2A}\right\rangle$	$\left\|Y_A\right\rangle^{IIa} = -0.713422\left\|X_0\right\rangle$ $+ \quad 0.131985\left\|X_{1A}\right\rangle + \quad 0.0904438$ $\left\|X_{2A}\right\rangle$	1.52849	0.282139	0.346174
IIb	$\left\|X_0\right\rangle,$ $\left\|X_{1A}\right\rangle,$ $\left\|X_{3A}\right\rangle$	$\left\|Y_A\right\rangle^{IIb} = 0.055315\left\|X_{1A}\right\rangle$ $+ 0.359176\left\|X_{1A}\right\rangle$ $+ 5.73182\cdot10^{-7}\left\|X_{3A}\right\rangle$	2.15865	0.451919	0.488893
IIc	$\left\|X_0\right\rangle,$ $\left\|X_{2A}\right\rangle,$ $\left\|X_{3A}\right\rangle$	$\left\|Y_A\right\rangle^{IIc} = 0.0132641\left\|X_0\right\rangle$ $+ 0.0618883\left\|X_{2A}\right\rangle$ $+ 4.14933\cdot10^{-7}\left\|X_{3A}\right\rangle$	1.82903	0.365689	0.414242
III	$\left\|X_0\right\rangle,$ $\left\|X_{1A}\right\rangle,\left\|X_{2A}\right\rangle,$ $\left\|X_{3A}\right\rangle$	$\left\|Y_A\right\rangle^{III} = 0.912459\left\|X_0\right\rangle$ $+ 0.776175\left\|X_{1A}\right\rangle - 0.209622\left\|X_{2A}\right\rangle$ $+ 9.70982\cdot10^{-7}\left\|X_{3A}\right\rangle$	2.36461	0.504184	0.53554

As a general observation, in all cases there was recorded a systematic increase of the correlation factor when computed in spectral space, that is, using the algebraic definition, as compared with the standard statistical values. We would like to take this opportunity to advocate the use of the algebraic definition instead of the statistical one since the first one has the physical meaning of the "length of action" respecting the old-fashioned dispersion analysis. Nevertheless, dispersion being a consequence of the appropriateness of the fit, the vectorial norm or the "length of action" accounts merely for the degree with which a certain model approaches the observed, or measured or manifested (chemical–biological) interaction. In this respect, it is also worth noting that both cationic and anionic predicted activities poorly resemble the experimentally expected activities with a smooth increase on the anionic influence, for all computed models except two cases based on the lipophylicity correlation (*Ia* and *IIa*).

Apparently, this behavior disagrees with the previously reported studies in which the anionic effect was only marginal in cytotoxicity [1, 4, 26], but in some special cases of certain ILs tests [7].

This situation was based on the observation that, for instance, since imidazolium ring in cations is a delocalized aromatic system with high electron acceptor potential, the nitrogen atoms are not capable to form any hydrogen bonds and the result is a very rigid and sterically inflexible system, while the elongation of R2 residue in side chain leads to a continuous increase of flexibility implying more conformational freedom [19]. The reason for this actions relies on the fragmental hydrophobicity of each carbon connected to a quaternary amine which, combining the geometric bond factor (that applies to the neutral solute) with a negative electronic bond factor, decreases its magnitude with the square of the distance from the central nitrogen atom [26, 27].

Moreover, the systematic variation of R1, R2, and so forth, at identical head groups and anions, in all published data, from the molecular to individual organism level, leads to the conclusion that the shorter the chain lengths of side chains the lower the cytotoxicity (higher EC50 values) [7, 9]. The electronic portion of the bond factor extends along a chain of no more than 5 alkane carbons causes a decrease in overall hydrophobicity so the chains longer than 5–6 atom carbons will have a greater permeability through the cell membrane (see for example [DMIM][BF4]) [26]. This, probably because the chemical transformation of the side chains of ILs may reduce toxicity as far as the metabolites are less toxic compared to their parent chemicals [7].

Next, aiming to combine the two somewhat separate effects of cations and anions in the IL activities, the SPECTRAL-SAR results are given in the Table 6.5, showing the working IL QSAR equation as well the actual vectorial norm, statistic and algebraic correlation factors for each mode of action envisaged so far.

Table 6.5. Spectral structure activity relationships (SPECTRAL-SAR) of the ionic liquids of Figure 6.2 against *Vibrio fischeri* toxicity, and the associated computed spectral norms with $\left\| |Y_{exp}\rangle \right\| = 4.41537$, statistic and algebraic correlation factors, computed upon the relations (6.6), (6.7), and (6.8), through all possible correlation models considered for the cationic-anionic interaction (6.11), respectively.

Mode	Ionic Liquid SPECTRAL-SAR	$\left\| \langle Y_{AC}\rangle^{Mode} \right\|$	$r_{S-SAR}^{STATISTIC}$	$r_{S-SAR}^{ALGEBRAIC}$
Ia	$\left| Y_{AC} \right\rangle^{Ia} = -0.36118 \left| X_0 \right\rangle$ $-0.124263 \left| X_{1C} \right\rangle + 0.291698 \left| X_{1A} \right\rangle$	2.58965	0.185959	0.586507
Ib	$\left| Y_{AC} \right\rangle^{Ib} = -0.603285 \left| X_0 \right\rangle$ $-0.0129369 \left| X_{2C} \right\rangle + 0.122745 \left| X_{2A} \right\rangle$	2.30825	0.179482	0.522776
Ic	$\left| Y_{AC} \right\rangle^{Ic} = 0.270054 \left| X_0 \right\rangle$ $+ 2.57881 \cdot 10^{-7} \left| X_{3C} \right\rangle$ $+ 5.11098 \cdot 10^{-7} \left| X_{3A} \right\rangle$	2.41575	0.259502	0.547123

Table 6.5. *(Continued)*

Mode	Ionic Liquid SPECTRAL-SAR	$\left\| \|Y_{AC}\rangle^{Mode} \right\|$	$r_{S-SAR}^{STATISTIC}$	$r_{S-SAR}^{ALGEBRAIC}$
IIa	$\|Y_{AC}\rangle^{IIa} = -0.908443\|X_0\rangle$ $-0.241666\|X_{1C}\rangle + 0.131985\|X_{2C}\rangle$ $+0.0295874\|X_{2C}\rangle +0.0904438\|X_{2A}\rangle$	2.88368	0.219986	0.653101
IIb	$\|Y_{AC}\rangle^{IIb} = -0.084473\|X_0\rangle$ $-0.222688\|X_{1C}\rangle + 0.359176\|X_{1A}\rangle$ $-1.62349\cdot10^{-6}\|X_{3A}\rangle + 5.73182\cdot10^{-7}\|X_{3A}\rangle$	3.48056	0.352356	0.788283
IIc	$\|Y_{AC}\rangle^{IIc} = 0.397721\|X_0\rangle$ $-0.124625\|X_{2C}\rangle + 0.0618883\|X_{2A}\rangle$ $-6.48599\cdot10^{-6}\|X_{3C}\rangle$ $+4.14933\cdot10^{-7}\|X_{3A}\rangle$	3.17318	0.299925	0.718667
III	$\|Y_{AC}\rangle^{III} = 1.05374\|X_0\rangle$ $-0.127251\|X_{1C}\rangle + 0.776175\|X_{1A}\rangle$ $-0.0677301\|X_{2C}\rangle - 0.209622\|X_{2A}\rangle$ $-4.48229\cdot10^{-6}\|X_{3C}\rangle$ $+9.70982\cdot10^{-7}\|X_{3A}\rangle$	3.7151	0.397148	0.841402

The data in Table 6.5 clearly demonstrate that the IL SPECTRAL-SAR models always predict higher norms in endpoint activities thus providing the considerable increase in the algebraic correlation factors with respect to the cationic and anionic subsystem effects. It is very interesting to see that the statistical values not only furnish lower values than the algebraic outputs for the ILs, but often lie even below the corresponding statistical values of the cationic and anionic subsystems. This situation gives us a chance to establish the limits of using the dispersion based correlation factor definition since it does not properly reproduced the addition effect of the two interacting subsystems as the anionic-cationic interaction in the IL structures. On the other hand, the mixture effect of the cationic and anionic vectorial actions in (eco)toxicological studies is well established as far as the single substances of a mixture acts in similar and close quantified manner in all considered modes.

The fact that this is the present case can be firstly visualized by the close inspection of Tables 6.3 and 6.4, where the spectral norms of the predicted activities feature close values in relatively narrow domain of actions. Moreover, the reinforcement of this idea comes from the data of Table 6.6 in which the angle of interactions between the cationic and anionic subsystems are computed, as given by the formula (6.15), for all considered endpoints with almost constant values around the 0.600 cosines of the angle, while the only higher fluctuation deviation appears in the *Ia* and *IIa* cases—the same previously evidenced for the cationic dominancy over the anion effects.

Table 6.6. The variation of the cosines of the anion-cationic correlation angle in vectorial space, according with the equation (6.15), for all considered modes of action of the ionic liquids of Figure 6.2 with the cationic and anionic subsystems SPECTRAL-SAR activities given in Tables 6.3 and 6.4, respectively.

Mode	Ia	Ib	Ic	IIa	IIb	IIc	*III*
$\cos\theta_{AC}$	0.66397	0.600124	0.562018	0.653248	0.60019	0.599635	0.591015

Overall, once the electronic and steric mechanisms have also been considered, the anion could play a central role as technicophore because it exhibits a high potential for change technological properties (solubility, viscosity) or due its peculiarity to partially decompose itself in the ion pair interaction [26, 28], leading to its dominant effect in the chemical–biological engaged activity.

With these consideration we can safely assume that the SPECTRAL-SAR in general and the algebraic correlation factors in particular are especially suited for modeling the (eco)toxicological activities of ILs from its anionic and cationic component effects.

The final part of discussion is devoted for picturing a mechanistically mode of action for cationic, anionic, and of their summed effects in containing ILs on the considered *Vibrio fischeri* species. That is, the minimum path procedure among all possible ways connecting endpoints from each category of models (i.e., with one, two, or three factors dependency) is to be considered. The path lengths were computed employing the equation (6.10) to all cationic, anionic and ILs data of Tables 6.3, 6.4, and 6.5, and the results are displayed in the Table 6.7, respectively. However, in order to identify the shortest paths in each category of endpoint connections, the following rules are applied: the first choice is the overall minimum path in a certain column of Table 6.7 (i.e., a system with a specific way of correlation factor); if the overall minimum belongs to many equivalent paths (as is the case of cationic with algebraic factor column in Table 6.7, for instance) the minimum path will be considered the one that links the starting endpoint with the closest endpoint in the sense of norms (as is the norm of *IIa* mode the closest to the norm of *Ia* mode in cationic-algebraic column of Table 6.7, for example); the overall minimum path will set the dominant hierarchy of the mechanistically mode of action towards the experimentally observed activity and will be called *the alpha path* (α); once the alpha path has been set the next minimum path will be looked for such that the starting endpoint should be different from the one involved in the alpha path (i.e., if in the alpha path the starting end point was *Ia*, the next path to be identified will begin either from the *Ib* or the *Ic* mode); the next minimum paths are chosen on the same rules as before and will be called as beta and gamma paths, β and γ, respectively.

At the end of this procedure each mode of action is "touched" only once, except for the final endpoint, here *III* (as a computational substitute of EXP), so that all the methodology being regarded as searching minimum path throughout variation of paths with the fixed final endpoint. Now, the alpha, beta, and gamma path can be easily identified in Table 6.7 and there are accordingly marked.

Table 6.7. Synopsis of the statistic and algebraic values of paths connecting the SPECTRAL-SAR models of Table 6.5, in the norm-correlation spectral-space of Figure 6.3, for the ionic liquids of Figure 6.2 against *Vibrio fischeri* toxicity. The primary, secondary, and tertiary—the so called alpha (α), beta (β), and gamma (γ) paths, are indicated according to the least path principle in spectral norm-correlation space with the statistic and algebraic variants of the correlation factors used, respectively.

Path	Value					
	Cationic		Anionic		Ionic Liquid	
	statistic	algebraic	statistic	algebraic	statistic	algebraic
Ia-IIa-III	0.324618	**0.320376 α**	1.0154	1.0049	1.14608	**1.15396 α**
Ia-IIb-III	0.324616	0.320376	**1.01536 γ**	**1.0049 γ**	**1.1451 α**	1.15396
Ia-IIc-III	**0.324616 α**	0.320376	1.01541	1.0049	1.14513	1.15396
Ib-IIa-III	0.739827	0.723082	**0.907864 β**	**0.899373 β**	**1.42694 γ**	1.44248
Ib-IIb-III	**0.739746 β**	0.723082	0.907871	0.899373	1.42377	**1.44248 γ**
Ib-IIc-III	0.739754	**0.723082 β**	0.907893	0.899373	1.42385	1.44248
Ic-IIa-III	**0.900418 γ**	0.86674	1.09986	1.08906	1.31968	1.33226
Ic-IIb-III	0.900057	**0.86674 γ**	0.630373	0.625533	1.30763	1.33226
Ic-IIc-III	0.90009	0.86674	**0.630371 α**	**0.625533 α**	**1.30908 β**	**1.33226 β**

With the aim of giving the complete results of the analysis in a single shot, Figure 6.3 is a "spectral" representation of the data, with the norms for each cationic, anionic, and resulted cationic liquid being drawn, linked by the major hierarchical paths against the considered mode of actions toward the observed "length" (norm) of the toxicity activity of the *Vibrio fischeri* species, for both statistical and algebraic correlation pictures. Figure 6.3 clearly underlines the fact that while anionic and cationic activity tendencies are somewhat complementary, they do not cancel each other in the IL that contains them but add up to attain the overall observed toxicity, in the spectral norm—correlation factor space. Besides this, some other useful information can be extracted from Figure 6.3 concerning the major path of structural causes in manifested toxicological action as revealed below.

While in cationic case the statistical and algebraic path does not coincide at all (e.g., what the alpha *Ia-IIc-III* path in statistic differs than the alpha *Ia-IIa-III* path in algebraic views), in the anionic case they are identically predicted (excepting the fact that in the algebraic case the shortened paths are registered), together providing a mixed behavior for the resulted IL (only the beta path *Ic-IIc-III* is overlapping between statistic and algebraic views). While in cationic and anionic subsystems the path hierarchies are reversed, as α→β→γ and γ→β→α, in the resulting IL again the mixture effect is recorded since the succession α→γ→β, against the successions of starting endpoints *Ia→Ib→Ic*, respectively. It is worth pointing out here that, indeed, the cationic alpha path is started on the lipophylicity causes (*Ia*), the same as for the containing IL, a different situation arising for the alpha anionic path that is beginning with the steric influence (*Ic*). This way the previously noted dominance of cationic influence when correlated with lipophylicity as well the recorded anionic influence related with steric effects are in this picture theoretically confirmed.

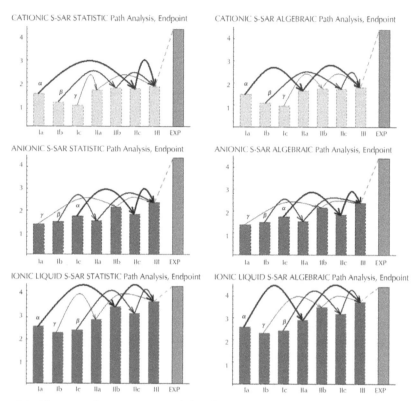

Figure 6.3. The spectral representation of the chemical–biological interaction paths across the SPECTRAL-SAR to the modeled endpoints of the *Vibrio fischeri*, according to the least (shortest) path rule within the spectral norm-correlation space applied on Table 6.7 data, for the cationic, anionic and resulted ionic liquid norms of Tables 6.3–6.5, for the statistic and algebraic versions of correlation factors, from up to down and left to right, respectively. The primary-alpha, secondary-beta, and tertiary-gamma path hierarchies of Table 6.7 are indicated by decreasing the thickness of the connecting lines.

Other interesting features about ecotoxicological paths rely on the fact that, while for cationic and anionic subsystems the algebraic paths are systematically lower that the corresponding statistical ones, in the IL case the situation is reversed. The interpretation of this fascinating result is that it also confirms that the chemical–biological IL dispersive (not specific) actions in environment are merely through its subsystem components than from itself as a whole. This result further motivates the design of ILs by tailoring the properties of its containing anionic and cationic components for prevent their hazardous toxicity.

Finally, let us also note that the ecotoxicological paths in cationic and anionic subsystems are summed up in the paths of the corresponding ILs in a not trivial ways:

$$\alpha_C + \gamma_A = \alpha_{AC}, \qquad (6.19)$$

$$\beta_C + \beta_A = \gamma_{AC}, \qquad (6.20)$$

$$\gamma_C + \alpha_A = \beta_{AC}. \qquad (6.21)$$

From these SPECTRAL-SAR equations some conceptual, however computationally based, ecotoxicological path rules for ILs can be concluded, namely:

- the anionic gamma path effect is marginal over the cationic alpha path—equation (6.19);
- the cationic and anionic beta paths decay into the gamma IL path when met together so that recording a sort of reciprocal cancellation of their effects—equation (6.20);
- the anionic alpha path effect is reinforcing over the cationic gamma path averaging both at the beta path level of the resulted IL—equation (6.21).

This way, the SPECTRAL-SAR model presented seems to provide a unitary picture of the anionic-cationic interaction in ILs as conciliating the anionic and cationic effects observed so far. However, further studies on different species with diverse computation schemes and parameters are required in order to conceptually asses a definitive theory of IL inter- and intra-mode of action.

CONCLUSIONS

Despite the promise to change "the face" of organic chemistry, the ILs are becoming a central issue in green chemistry as well, due to their unique physico-chemical properties. For instance, their high viscosities compared with conventional molecular solvents have impact on lowering the reaction rates while through their higher degree of immersion into ground soil and sediments the ecotoxicity of various species and organisms are recorded. The latter effects are based on their common supposed mechanism through membrane disruption: having structural similarity with detergents, pesticides, and antibiotics [4, 20, 29, 30] they induce polar narcosis due to their interfacial properties and may cause membrane-bond protein disruption [2, 24]. ILs modes of action are based on the fact that the disrupt membranes and hydrophobic molecules have a greater ability to accumulate at this interface [9], while other mechanisms may arise from acetylcholinesterase inhibition [31], from common cellular structure or processes [4], from structural DNA damage [23, 32]. Eventually, some bacteria could potentially break down imidazolium into different metabolites which follow further inhibition on themselves [1]. In any case, low lipophylicity imidazolium based ILs values indicate low permeability of these ILs [2].

In addition, to the certain role that the lipophylicity has in the assessment of IL toxicity on various species, before constructing kit- or battery-tests involving chemical–biological level of organization, from enzyme to cellular to short and long lived organism [2, 4, 7–9, 23, 24, 26, 31–44], worth focusing on the quantitative algorithm with the help of which the experimental data will be interpret. This is because, criticized for its lacking in real ecological meaning, the dose-response approach for estimating the lethal effects of toxicants on organisms regulatory norms have been built around LC50 values that can be compared between different toxicants and organisms [2, 45], while the statistical correlation factors of the proposed QSARs so far are based on dispersion (i.e., hazardous) action and less accounting on the specific ways of ILs action. In this regard, it is important to know if the tested chemicals could reach the target site into organism in a specifically manner first, so that an extrapolation from *in vitro* to *in vivo* is necessary [31] while other factors like biodegradation and bioaccumulation are necessarily be considered before conclusions are drawn [24, 46, 47].

Having to consider both specific and hazardous paths of action concerning the environmental risk posed by the newly released ILs, the recently introduced SPEC-TRAL-SAR scheme of quantification of chemical–biological interaction [18] is adapted here to the particular IL cationic-anionic structural interaction.

The present method is based on the previous evaluation of QSARs with specific norms and algebraic correlation factors for cationic and anionic influences followed by their vectorial interference in the spectral norm space. With a short generation lifetime, bacteria is an ideal starting point for IL SAR investigation and serves as a basis for further toxicity tests to higher organisms and more complex systems.

This way, the SPECTRAL-SAR-IL model adapted here was applied to *Vibrio fischeri* species, this species being reported as the most resistant to other conventional environmental releasing chemical compounds [21]. The results are promising, confirming at some level of analysis all previous certified facts, the cationic generally dominance [48], the special anionic effects [49], the ionic pair effect [50], and so forth, adding in an elegant manner the least (or minimum) path rule in the spectral norm-correlation factor space as an effective tool in which the ecotoxicological rules of the ILs can be derived.

Worth, however, noted that when choosing the Hansch factors, although there was already pointed out that the use of free enthalpy instead of total energy provides better statistical correlation factors [51], in the present study the total energy was preferred due to its close connection with the sterical effects through the computational geometrical optimization procedure involved. This way, the minimal comprehensive set of QSAR descriptors, that is hydrophobic, electronic, and steric, when predict the ecotoxicity endpoints was assured [21]. In this respect, further studies may address the effect of the length of alkyl chain of the imidazolium-based IL on the activity.

The present approach leaves room for further investigation at both the conceptual and computational levels of expertise when improving the specificity of analysis in terms of considered structural factors and of their SAR combinations or when generalizing it to the eco-design of multiple species battery, respectively.

KEYWORDS

- **Acetylcholinesterase**
- **Biodegradability**
- **Cytotoxicity**
- **Lipophylicity**

Chapter 7

SPECTRAL-SAR Ecotoxicology of Ionic Liquids: The *Daphnia magna* Case

Mihai V. Putz, Ana-Maria Putz, and Vasile Ostafe

INTRODUCTION

Aiming to provide an unified theory of ionic liquids' ecotoxicity the recent spectral structure-activity relationship (SPECTRAL-SAR) algorithm is employed for testing the two additive models of ionic liquid activity: the parametric (abbreviated as $|0+\rangle$) and the endpoint (abbreviated as $|1+\rangle$) models accounting for the causal and observed interactions between the anionic and cationic subsystems, respectively. As a working system the *Daphnia magna* ecotoxicity was characterized through the formulated and applied spectral chemical–biological interaction principles. Specific anionic-cationic rules of interaction along the developed mechanistic hypersurface map of the main ecotoxicity paths together with the so called resonance limitation of the standard statistical correlation analysis were revealed.

Since the reformulation of the classical quantitative structure-activity relationship (QSAR) modeling under the SPECTRAL-SAR analytical analysis [1–3], also the ecotoxicological studies have been reinforced with new tools linking the molecular structure of the chemicals dispersed in environment over certain species with their recorded biological activities [4–5].

Basically, the applied SPECTRAL-SAR algorithm can be summarized by means of the so called SPECTRAL-SAR operator defined as a collection of successive operations,

$$\hat{O}_{S-SAR}: \begin{cases} Det\left(|Y\rangle, |X_0\rangle, |X_1\rangle, ..., |X_M\rangle\right) = 0, \||Y\rangle\|, r_{S-SAR}^{ALGEBRAIC}, \\ \delta\left[A_{(\bullet|\bullet)}, B_{(\bullet|\bullet)}\right] = 0, \langle\alpha, \beta, \gamma, ...\rangle, \quad \begin{array}{l} A, B: ENDPOINTS \\ \alpha, \beta, \gamma, ...: SPECTRAL\ PATHS \end{array} \end{cases} \quad (7.1)$$

employing the N-structural molecular input data, as the vectors $|X_0\rangle = |1_1 ... 1_N\rangle$, $|X_i\rangle$, $i=1, ... M$, with, with the predicted endpoints $|Y\rangle$, via the scalar product based spectral norm $\||Y\rangle\|$, by using algebraic correlation factor $r_{S-SAR}^{ALGEBRAIC}$ as the ratio of calculated to the measured spectral norm, until the first minimum paths across the spectral norms-correlation factors are attained between the specific endpoints [3]. Remarkably, the above described analytical steps of a SPECTRAL-SAR analysis can be transposed

Putz M.V., Lacrămă A.M., Ostafe V. "Spectral-SAR Ecotoxicology of Ionic Liquids. The *Daphnia magna* Case", *International Journal of Ecology (former Research Letters in Ecology)*, ISSN: 1687-9708 e-ISSN: 1687-9716, Article ID12813/5 pages, DOI: 10.1155/2007/12813, 2007.

into the driving principles of the associated spectral chemical eco- and bio-studies. They can be, respectively, formulated as:

- *Principle 1* states that the "orthogonality" of assumed molecular factors which correlate with eco- and bio-effects is assured by the spectral decomposition of the associate activity respecting them;
- *Principle 2* states that the "length" of the predicted (eco)biological action follows the self-scalar product rule of the computed endpoint activity;
- *Principle 3* states that the "intensity" of the chemical-eco-bio-interaction is determined by the ratio of the expected to measured activity norms;
- *Principle 4* states that the "selection" of the manifested chemical-eco-(bio-) binding parallels the minimum distances of paths connecting all possible endpoints in the norm-correlation hyperspace;
- *Principle 5* states that the "validation" of the obtained mechanistic picture is done by constraining that the influential minimum paths are numbered by the cardinal of the input structural factors set so that, excepting the final endpoint which is always considered as the final evolution target, all other endpoints are activated one and one time only.

With these concepts and principles the chemical-eco-biological phenomenology can be mechanistically unfolded such that from all virtual theoretical possibilities connecting molecular structure with the observed effects that ones which are most influential will be systematically discovered.

In the present study, the above algorithm is applied to the intriguing case of ionic liquids acting on the eco-paradigmatic *Daphnia magna* species within two different additive models of the Hansch expansion.

HANSCH SPECTRAL-SAR-IONIC LIQUID ECOTOXICOLOGICAL /0+> AND /1+> MODELS

Usually, when considering Hansch QSAR expansion the hydrophobic, electronic, and steric factors have to be considered. While the hydrophobicity index *LogP* describes at the best the quality of molecular transport through cellular membranes, for the electronic and steric contributions many structural parameters may be considered [3]. However, we consider that the polarizability (*POL*) measures the inductive electronic effect reflecting the long range or van der Waals bonding whereas for the steric component the total energy (E_{TOT}) is assumed as the representative index since it is calculated at the optimum molecular geometry where the stereo-specificity is included. The recent ecotoxicological studies based on these chemical descriptors have proved their reliability in providing the molecular mechanism according which chemicals act upon certain species [3–5].

Nevertheless, when there is about of the ionic liquids (*IL*) eco-bio-logical influence the particular anionic-cationic structure has to be properly considered since almost all ionic liquids structural information root on the superposition of the separate anionic and cationic contributions. In this situation, two different additive models for modeling anionic-cationic interaction can be considered.

The first one is based on the vectorial summation of the produced anionic and cationic biological effects. In other words, this so called $|1+\rangle$ model is constructed on the superposition of the anionic (subscripted with A) and cationic (subscripted with C) activities and can be formally represented as:

$$|Y_{AC}\rangle^{1+} = |1+\rangle = |Y_A\rangle + |Y_C\rangle = \hat{O}_{S-SAR}\left[g\left(\{|X_A\rangle\}\right) + g\left(\{|X_C\rangle\}\right)\right] \qquad (7.2a)$$

with Hansch combinations

$$\{|X_{A,C}\rangle\} = \{LogP_{A,C}, POL_{A,C}, E_{TOT(A,C)}\}. \qquad (7.2b)$$

Practically, with the $|1+\rangle$ model the SPECTRAL-SAR procedure encompassed by the operator (7.1) is performed for anionic and cationic subsystems separately, and then added up in the resulting IL-activity. The analysis based on this model was recently reported for the *Vibrio fischeri* ecotoxicity [4]. However, care must be taken on the "angle" between the anionic and cationic vectors that can control the final IL-vectorial resultant. In principle, there is possible that in the case of almost overlapping of anionic and cationic bio- or eco-toxic activities their effect to be in "resonance" so that to predict a larger effect than that observed, a situation clearly indicating a further self-potential of action, in a given ecosystem.

The second SPECTRAL-SAR model can be advanced here when the additive stage is considered at the incipient stage of the SPECTRAL-SAR operator (7.1), so that the considered Hansch factors are firstly combined to produce the anionic-cationic (subscripted with AC) indices that are further used to produce the spectral mechanistic map of the concerned interaction. This way, the so called $|0+\rangle$ model is produced:

$$|Y_{AC}\rangle^{0+} = \hat{O}_{S-SAR}|0+\rangle = \hat{O}_{S-SAR}\, f\left(\{|X_A\rangle\}, \{|X_C\rangle\}\right) \qquad (7.3a)$$

with the Hansch specification of the spectral vectors:

$$f\left(LogP_A, LogP_C\right) \equiv LogP_{AC} = \log\left(e^{LogP_A} + e^{LogP_C}\right) \in \{|X_{1AC}\rangle\}, \qquad (7.3b)$$

$$f\left(POL_A, POL_C\right) \equiv POL_{AC} = \left(POL_A^{1/3} + POL_C^{1/3}\right)^3 \in \{|X_{2AC}\rangle\}\ [\text{Å}^3], \qquad (7.3c)$$

$$f\left(E_A, E_C\right) \equiv E_{AC} = E_A + E_C - 627.71\frac{q_A q_C}{POL_{AC}^{1/3}} \in \{|X_{3AC}\rangle\}\ [\text{kcal/mol}]. \qquad (7.3d)$$

The open issue addresses whether the $|0+\rangle$ and $|1+\rangle$ states leaves with the same results or in which aspects of the SPECTRAL-SAR operator (7.1) they might differ in the IL ecotoxicity. In this regard the present *Daphnia magna* case is founded most suggestive, with the analytical results presented in what next.

RESULTS OF IONIC LIQUID ACTIVITY ON *DAPHNIA MAGNA* SPECIES

Daphnia magna [6, 7], Figure 7.1, is a pelagic freshwater crustaceans standing as one of the established species for the standard toxicity test organism because of easiness to culture it in laboratory being also sensitive to a variety of pollutants [8, 9], while measuring both acute and chronic effects for the survival and reproductive parameters, respectively [10]. It can be used also as an important link between microbial and higher trophic levels, furnishing the main environment for hundreds of physiological, evolutionary, and ecological studies [9, 11].

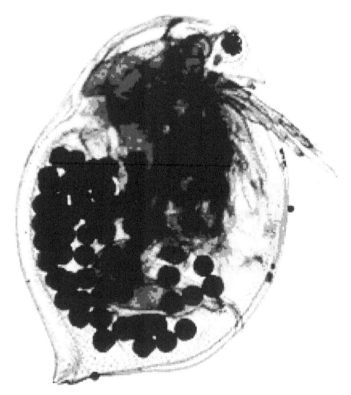

Figure 7.1. Daphnia magna with eggs [7].

However, although there were developed many tests using *Daphnia magna*, being considered, for instance, as a sub-lethal indicative parameter in aquatic toxicity tests with lactate dehydrogenase activity [12], as a biomarker for sub-lethal toxicity assessment of chemical—contaminated groundwater when haem peroxidase activity is measured [10], for the assessment of the potential impact of new chemicals on the aquatic environment [13, 14], for the leaching toxicity of metal containing solid wastes using 24 h immobilization test [15, 16], or by quantifying endocrine disrupters toxicity by means of acute toxicity tests [17], the measured ionic liquids activity remains unexplained in terms of molecular ways of action. To complete this, the above SPECTRAL-SAR ecotoxicological principles are systematically applied and interpreted for the observed toxicity of the ionic liquids of Table 7.1, with either $|0+>$ and $|1+>$ models, see equations (7.3) and (7.2), respectively.

Table 7.1. The studied ionic liquids actions on *Daphnia magna* species with the toxic activities A_{exp} = $Log(EC_{50})$ [9], while the marked values were taken from [18], along structural parameters *LogP, POL*, and E_{TOT} as accounting for the hydrophobicity, electronic (polarizability), and steric (total energy at optimized 3D geometry) effects, computed with HyperChem program [19], for each cation and anion fragment, as well as for the anionic-cationic |0+> composed state, by means of equations (7.3b)–(7.3d), respectively.

Ionic Liquid Compound		A_{exp}	LogP			POL [Å³]			E_{TOT} [kcal/mol]		
Structure	Name	$|Y_{exp}>$	$|X_{1C}>$	$|X_{1A}>$	$|X_{1AC}>$	$|X_{2C}>$	$|X_{2A}>$	$|X_{2AC}>$	$|X_{3C}>$	$|X_{3A}>$	$|X_{3AC}>$
	1-n-octyl-3-methylpyridinium bromide	– 2.60	4.90	0.94	4.92	26.69	3.01	87.08	–371060.81	–1596918.25	–1967840
	1-n-hexyl-3-methylpyridinium bromide	–2.41	4.11	0.94	4.15	23.02	3.01	78.87	–322641.81	–1596918.25	–1919410
	1-n-butyl-3-methylpyridinium bromide	–1.24	3.32	0.94	3.41	19.35	3.01	70.37	–274222.62	–1596918.25	–1870990
	1-n-octyl-3-methylimidazolium bromide	–4.33	2.26	0.94	2.5	24.56	3.01	82.35	–357484.59	–1596918.25	–1954260
	1-n-hexyl-3-methylimidazolium bromide	–2.22	1.47	0.94	1.93	20.89	3.01	73.98	–309065.84	–1596918.25	–1905830
	1-n-butyl-3,5-dimethylpyridinium bromide	–1.01	3.78	0.94	3.84	21.18	3.01	74.65	–298437.03	–1596918.25	–1895210
	1-n-hexyl-4-piperidino pyridinium bromide	–3.66	4.63	0.94	4.65	30.93	3.01	96.25	–452857.03	–1596918.25	–2049640
	1-n-hexyl-4-dimethylamino pyridinium bromide	–3.28	3.91	0.94	3.96	26.2	3.01	86.00	–380945.12	–1596918.25	–1977720
	1-n-hexyl-3-methyl-4-dimethylamino pyridinium bromide	–2.79	4.37	0.94	4.40	28.04	3.01	90.03	–405145.97	–1596918.25	–2001920
	1-n-hexylpyridinium bromide	–1.93	3.64	0.94	3.71	21.18	3.01	74.65	–298427.37	–1596918.25	–1895200

Table 7.1. *(Continued)*

Ionic Liquid Compound		A_{exp}	LogP			POL [Å³]				E_{TOT} [kcal/mol]	
Structure	Name	$|Y_{exp}>$	$|X_{1C}>$	$|X_{1A}>$	$|X_{1AC}>$	$|X_{2C}>$	$|X_{2A}>$	$|X_{2AC}>$	$|X_{3C}>$	$|X_{3A}>$	$|X_{3AC}>$
	1-n-hexyl-2,3-dimethylimidazolium bromide	−2.19	1.67	0.94	2.06	22.72	3.01	78.19	−333284.94	−1596918.25	−1930060
	1-n-butyl-3-methylimidazolium chloride	−1.07*	0.68	0.63	1.34	17.22	2.32	59.60	−260646.64	−285190.78	−545677
	1-n-butyl-3-methylimidazolium bromide	−1.43*	0.68	0.94	1.51	17.22	3.01	65.26	−260646.64	−1596918.25	−1857410
	1-n-butyl-3-me-thylimidazolium tetrafluoroborate	−1.32*	0.68	1.37	1.78	17.22	2.46	60.80	−260646.64	−261310.59	−521798
	1-n-butyl-3-me-thylimidazolium hexafluorophosphate	−1.15*	0.68	2.06	2.28	17.22	1.78	54.62	−260646.64	−580264.94	−840746
	Tetrabutyl ammonium bromide	−1.53	4.51	0.94	4.54	30.91	3.01	96.21	−422421.97	−1596918.25	−2019200
	Tetrabutyl phosphonium bromide	−2.05	2.89	0.94	3.02	30.91	3.01	96.21	−600149.625	−1596918.25	−2196930

In such, when considering the principles of orthogonality, of the length and of intensity of toxicological actions the Table 7.2 is produced, where, for comparison, the standard statistical factor was also added. There can be first observed that both anionic and cationic fragments have quite important contribution to the "length" and "intensity" of ionic liquids ecotoxicity through the computed spectral norms and algebraic correlation factors, respectively, very close to the experimental one, that is to 9.59481. Then, in all cases, the mode of action in which all three Hansch factors were considered (mode *III* with $LogP+POL+E_{TOT}$) records the best norm and correlations being the closest description of the ionic liquids-*Daphnia magna* chemical–biological interaction. As well, the cationic influence is found with the dominant contribution over the anionic effects in ecotoxicity, in all considered Hansch modes of action.

Nevertheless, the statistical correlation factors always yields lower values than the corresponding algebraically ones. Moreover, with the $|1+>$ model there are even recorded imaginary statistical correlations of the computed endpoints $|Y_{AC-Mode}>^{1+}$ whereas the algebraically outputs give almost the sum of the anionic and cationic length and intensity endpoint activity. This can be phenomenologically explained by the so called resonance effect when almost zero angles between the anionic and cationic endpoint vectors, for all molecular modes of actions of Table 7.2, are obtained as clearly evidenced by the cosines values of Table 7.3 [4].

Table 7.2. Spectral structure-activity relationships (SPECTRAL-SAR) of the ionic liquids toxicity of Table 7.1 against *Daphnia magna* species, and the associated computed spectral norms, with $\||Y_{EXP}>\|=9.59481$, statistic and algebraic correlation factors [3, 4], throughout the possible correlation models considered from the *anionic, cationic,* and *ionic liquid* $|1+>$ and $|0+>$ states in Table 7.1, respectively.

Mode	SPECTRAL-SAR Equations	$\||Y\rangle^{Mode}\|$	$r_{S-SAR}^{STATISTIC}$	$r_{S-SAR}^{ALGEBRAIC}$
Ia	$|Y_{A-Ia}>=-2.91325|X_0>+0.773243|X_{1A}>$	8.83127	0.266552	0.920421
	$|Y_{C-Ia}>=-1.41993|X_0>-0.250543|X_{1C}>$	8.92169	0.420761	0.929845
	$|Y_{AC-Ia}>^{0+}=-1.21571|X_0>-0.287831|X_{1AC}>$	8.89048	0.374616	0.926593
	$|Y_{AC-Ia}>^{1+}=|Y_{A-Ia}>+|Y_{C-Ia}>$	17.6883	2.21964 i	1.84353
Ib	$|Y_{A-Ib}>=1.40126|X_0>-1.23268|X_{2A}>$	8.94784	0.455964	0.932572
	$|Y_{C-Ib}>=0.502967|X_0>-0.113186|X_{2C}>$	9.06691	0.59121	0.944981
	$|Y_{AC-Ib}>^{0+}=1.35675|X_0>-0.0447316|X_{2AC}>$	9.08979	0.613973	0.947366
	$|Y_{AC-Ib}>^{1+}=|Y_{A-Ib}>+|Y_{C-Ib}>$	17.9079	2.19638 i	1.86641
Ic	$|Y_{A-Ic}>=-0.851715|X_0>+9.25362\cdot10^{-7}|X_{3A}>$	8.96309	0.475327	0.934161
	$|Y_{C-Ic}>=-0.426598|X_0>+4.93426\cdot10^{-6}|X_{3C}>$	8.95817	0.469161	0.933648
	$|Y_{AC-Ic}>^{0+}=-0.555261|X_0>+9.12119\cdot10^{-7}|X_{3AC}>$	8.99267	0.510889	0.937244
	$|Y_{AC-Ic}>^{1+}=|Y_{A-Ic}>+|Y_{C-Ic}>$	17.8233	2.20167 i	1.8576
IIa	$|Y_{A-IIa}>=3.01778|X_0>-0.572955|X_{1A}>-1.59437|X_{2A}>$	8.96021	0.47173	0.933861
	$|Y_{C-IIa}>=0.624239|X_0>+0.0453509|X_{1C}>-0.123924|X_{2C}>$	9.06885	0.59317	0.945183
	$|Y_{AC-IIa}>^{0+}=1.63496|X_0>+0.138794|X_{1AC}>-0.0539568|X_{2AC}>$	9.1014	0.62522	0.948575
	$|Y_{AC-IIa}>^{1+}=|Y_{A-IIa}>+|Y_{C-IIa}>$	17.9161	2.1931 i	1.86727
IIb	$|Y_{A-IIb}>=-1.03967|X_0>+0.134654|X_{1A}>+8.88035\cdot10^{-7}|X_{3A}>$	8.96426	0.476781	0.934283
	$|Y_{C-IIb}>=-0.468198|X_0>-0.143221|X_{1C}>+3.63796\cdot10^{-6}|X_{3C}>$	8.99112	0.50908	0.937082
	$|Y_{AC-IIb}>^{0+}=-0.48744|X_0>-0.0777712|X_{1AC}>+8.08314\cdot10^{-7}|X_{3AC}>$	8.99774	0.51675	0.937772
	$|Y_{AC-IIb}>^{1+}=|Y_{A-IIb}>+|Y_{C-IIb}>$	17.8793	2.21324 i	1.86343
IIc	$|Y_{A-IIc}>=0.0383859|X_0>-0.445991|X_{2A}>+6.44823\cdot10^{-7}|X_{3A}>$	8.96808	0.481497	0.93468
	$|Y_{C-IIc}>=0.657036|X_0>-0.185992|X_{2C}>-4.45973\cdot10^{-6}|X_{3C}>$	9.09155	0.615686	0.947549
	$|Y_{AC-IIc}>^{0+}=1.26793|X_0>-0.040945|X_{2AC}>+1.19517\cdot10^{-7}|X_{3AC}>$	9.09116	0.615307	0.947508
	$|Y_{AC-IIc}>^{1+}=|Y_{A-IIc}>+|Y_{C-IIc}>$	17.9586	2.19937 i	1.8717

Table 7.2. *(Continued)*

Mode	SPECTRAL-SAR Equations	$\left\|Y\right)^{Mode}\right\|$	$r_{S-SAR}^{STATISTIC}$	$r_{S-SAR}^{ALGEBRAIC}$
III	$\|Y_{A-III}\rangle = 0.893032\|X_0\rangle - 0.218536\|X_{1A}\rangle -$ $0.721371\|X_{2A}\rangle + 5.32183 \cdot 10^{-7}\|X_{3A}\rangle$	8.96926	0.482946	0.934803
	$\|Y_{C-III}\rangle = 1.38233\|X_0\rangle + 0.223392\|X_{1C}\rangle - 0.299344\|X_{2C}\rangle -$ $8.16291 \cdot 10^{-6}\|X_{3C}\rangle$	9.12145	0.644232	0.950666
	$\|Y_{AC-III}\rangle^{0+} = 1.53849\|X_0\rangle + 0.141501\|X_{1AC}\rangle -$ $0.0497924\|X_{2AC}\rangle + 1.37119 \cdot 10^{-7}\|X_{3AC}\rangle$	9.10319	0.62694	0.948762
	$\|Y_{AC-III}\rangle^{1+} = \|Y_{A-III}\rangle + \|Y_{C-III}\rangle$	17.9531	2.17933 i	1.87113

Table 7.3. The values of the cosines of the anion-cationic vectorial angles [4], for all considered modes of action of Table 7.2, for the $|1+\rangle$ states of the ionic liquids of Table 7.1.

Mode	Ia	Ib	Ic	IIa	IIb	IIc	III
$\cos\theta_{AC}$	0.985468	0.976338	0.978196	0.975018	0.983081	0.97768	0.969683

Instead, within $|0+\rangle$ model all the lengths and intensities of the endpoints $|Y_{AC-Mode}\rangle^{>0+}$ approach a kind of average of anionic and cationic ecotoxicological effects with a smooth increase over the individual cationic effects for the modes *Ib* (*POL*), *Ic* (E_{TOT}), *IIa* (*LogP+POL*), and *IIb* (*LogP+E_{TOT}*). However, to further decide which of these modes is further selected by the binding mechanism, the remaining spectral ecotoxicological principles, namely the selection and validation principles are finally employed with the results collected in Table 7.4.

Table 7.4. Synopsis of the statistic and algebraic values of the paths connecting the SPECTRAL-SAR models of Table 7.2, in the norm-correlation spectral-space, for *Daphnia magna* species against the ionic liquids toxicity of Table 7.1; the primary, secondary, and tertiary—the so called alpha (α), beta (β), and gamma (γ) paths, are indicated according to the "selection" and "validation" principles in norm-correlation spectral space when the statistic and algebraic variants of the correlation factors are respectively used.

Path	Cationic		Anionic		Ionic Liquid			
					state $\|0+\rangle$		state $\|1+\rangle$	
	statistic	algebraic	statistic	algebraic	statistic	algebraic	statistic	algebraic
Ia-IIa-III	0.299988	0.200851	0.256742 γ	0.13874	0.330033	0.213862 γ	0.260535	0.266181
Ia-IIb-III	0.300103 γ	0.200851	0.2567	0.13874	0.330581 γ	0.213862	0.25639 γ	0.266181 γ
Ia-IIc-III	0.299895	0.200851 γ	0.25666	0.13874 γ	0.33011	0.213862	0.269477+ R* i	0.277223
Ib-IIa-III	0.07607 α	0.0548409 α	0.034447	0.0215298 β	0.0186427	0.0134673	0.0418672 α	0.0454552 α
Ib-IIb-III	0.299514	0.207241	0.0344468	0.0215298	0.286398	0.198562	0.0886683	0.102966
Ib-IIc-III	0.0760732	0.0548409	0.0344464 β	0.0215298	0.0186427 α	0.0134673 α	0.0506137+ R* i	0.0564973
Ic-IIa-III	0.23953	0.16417	0.0190146	0.0119873	0.160257 β	0.111113	0.126723	0.130484
Ic-IIb-III	0.23952	0.16417 β	0.00980164 α	0.00619966 α	0.160264	0.111113 β	0.120323 β	0.130484 β
Ic-IIc-III	0.239484 β	0.16417	0.00980164	0.00619966	0.16027	0.111113	0.135252+ R* i	0.141526

*R = 0.0192723

As a note, there is interesting that although different in their analytical formulations, the additive parametric and endpoint models, $|0+\rangle$ and $|1+\rangle$, furnish the same hierarchies of the paths for the chemical–biological actions, see Table 7.4. Worth mentioning that the statistical imaginary correlation values for the ionic liquids $|1+>$ states in Table 7.2 extend their behavior in the Table 7.4 too; however, these paths are avoided from the validation principle. In other words, for instance, if the alpha (α) path starts with *Ib* (on *POL*) molecular mode of action in the ionic liquid $|0+>$ state, the alpha path in ionic liquid $|1+>$ state begins with the same molecular mode of action even following different intermediate mode until the common final endpoint (the *III* mode). The same happens also with the beta (β) and gamma (γ) paths of considered ionic liquids-*Daphnia magna* ecotoxicity. However, this rule is not met at the cationic and anionic fragments, while the dominant cationic effects can be noted also at the least paths level since the nature of cationic mechanism is preserved to the ionic liquids nature according with the spectral path equations:

$$\alpha_C + \beta_A = \alpha_{IL} \tag{7.4a}$$

$$\beta_C + \alpha_A = \beta_{IL} \tag{7.4b}$$

$$\gamma_C + \gamma_A = \gamma_{IL} \tag{7.4c}$$

Finally, the results of all SPECTRAL-SAR ecotoxicological principles applied to ionic liquids-*Daphnia magna* case of chemical-eco-biological interaction can be unitarily presented in the Figure 7.2, where the spectral hypersurface was generated by the 3D interpolation of all lengths (norms) for all the endpoint modes of Table 7.2, for all cationic, anionic, $|0+>$, and $|1+>$ states of ionic liquids of Table 7.1.

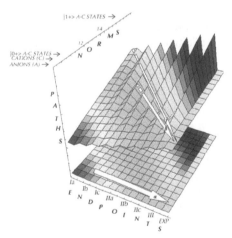

Figure 7.2. The spectral hypersurface of the structural hierarchical paths toward the recorded ecotoxicological *(EXP)* activity (in the extreme right hypersurface region) of the ionic liquids of Table 7.1 on *Daphnia magna* species: the alpha path (α) initiates on the polarizability *(Ib)* anionic-cationic interaction (in the left-bottom hypersurface region), being followed by the beta path (β) which originates on the steric *(Ic)* anionic-cationic interaction (in the left-top hypersurface region hypersurface region), and successively by the gamma path (γ) based on the hydrophobic *(Ia)* anionic-cationic interaction (in the extreme left-top hypersurface region) of the norm-correlation spectral space of Table 7.2, with the decaying order of the thickness of the connecting arrows, respectively.

The alpha dominant paths are easily identified, according with the Table 7.4, as originating in the *Ib*, that is on *POL*arizability or van der Waals molecular mode of action, while the beta and gamma ones starts with the steric (*Ic*: E_{TOT}) and hydrophobic (*Ia*: *Log*P) specific chemical–biological binding, respectively. Further studies may address other Hansch, as well as other topological and quantum molecular factors to further test and use the present spectral ecotoxicological principles on batteries with different species.

CONCLUSION

In the context of QSAR ecotoxicological studies the SPECTRAL-SAR model is firstly employed to produce the so called spectral ecotoxicological principles. They are able to offer the complete picture of molecular specific interaction between a series of chemicals with certain species by generating the molecular mechanisms of actions; as well they are opening room for preventing and controlling the envisaged bio- and eco-logical systems. The particular case of *Daphnia magna* was presented respecting a selected series of 17 ionic liquid ecotoxicity, within the introduced cationic-anionic Hansch models by the additive factors and endpoints, the $|0+\rangle$ and $|1+\rangle$ SPECTRAL-SAR-IL states, respectively. The application of the spectral ecotoxicological principles on this chemical–biological coupled system revealed that the cationic fragments clearly dominate the anionic effects by driving the containing ionic liquid specific interaction paths. Moreover, it was found that the SPECTRAL-SAR operator may identically transform the $|0+\rangle$ state into $|1+\rangle$ respecting the origin of specific interaction while producing different intermediate paths through the recorded endpoints:

$$\hat{O}_{S-SAR}|0+\rangle \begin{cases} =|1+\rangle, & \text{fixed endpoints} \\ \neq|1+\rangle, & \text{path endpoints} \end{cases}. \tag{7.5}$$

The resonance effect, manifested when the anionic and cationic endpoint vectors are almost parallel in the additive $|1+\rangle$ state of ionic liquids, was also met for the *Daphnia magna* case, leading with the lesson that in such case the other additive model, the parametric additive $|0+\rangle$ state of ionic liquids, should be employed, avoiding the associate imaginary statistical correlations and activity paths.

Overall, it resulted that the primary path of bonding between the working ionic liquids and *Daphnia magna* species occurs via molecular polarizability, thus emphasizing on the ionic or on the long range chemical–biological interaction, followed by the steric and the hydrophobic Hansch mechanisms through the beta and gamma manifested paths. Nevertheless, the presented methodology leaves with the possibility of analytical characterization of bio-and eco-activity of other species against given set of trained or new synthesized chemicals, as well as for the inter-species correlations.

KEYWORDS

- ***Daphnia magna***
- **Ecotoxicological principles**
- **Hansch factors**
- **Polarizability**

Chapter 8

SPECTRAL-SAR Ecotoxicology of Ionic Liquids: The *Electrophorus electricus* Case

Mihai V. Putz, Ana-Maria Putz, Vasile Ostafe, and Adrian Chiriac

INTRODUCTION

The spectral structure-activity relationship (SPECTRAL-SAR) method, which re-places the old-fashioned multi-regression one, is an exclusive algebraic way of quan-titative structure-activity relationship model; it provides endpoints hierarchy scheme that opens the perspective to further design ecotoxicological test batteries from the molecular to cellular, organism level, and finally multi-species test batteries. How-ever, when about ionic liquid (IL) modeling, two additive models of anionic-cationic interaction containing IL activity, that is the so called $|0+>$ and $|1+>$ models, are to be respectively tested for better assessment of mechanistic models toward chemical–bio-logical interaction. The present studies, reviewed both this methods on an application upon the acethylcolinesterase by employing its reported ecotoxicological activities among some ILs with different cations (imidazolium, phosphonium, pyridinium) and with different anions (e.g., BF4, PF6, Cl, Br).

Within the last years, electric cells of *Electrophorus* [1] have provided a unique model system that is both specialized and appropriate for the study of excitable cell membrane electrophysiology and biochemistry. Electric tissue generates whole animal electrical discharges by means of membrane potentials that are distinctively similar to those of mammalian neurons, myocytes, and secretory cells [2].

On the other side, acetylcholinesterase (EC 3.1.1.7) enzyme is an essential part of the nervous system of almost all-higher organisms, including man. Because of the reason that the active center of the enzyme is very conservative among organisms, it is allowed the extrapolation of structure-activity relationships between organisms [3]. Nowadays, we confront with the intensification of the EU chemicals legislation (REACH project) so, also with the interest of the ecotoxicological hazard potential of chemicals [4]; generally, the lipophilicity of chemical substances is considered to be the mediator of non-specific toxic effects, called by membrane interactions, which are named: *narcosis* (polar and non-polar (when applied to mammals) or *baseline toxic-ity* (in aquatic toxicity) [4]. When about environment disposal of solvents, ILs, due to their promising physical and chemical properties, are now becoming the central issue in ecotoxicologically studies [5]. One of the most interesting applications of ILs is the possibility to design and synthesize new solvents with properties tailored for specific applications [6], and also with minimal toxicity to the environment [7]. For the design-ing of different ILs, hydrophobicity and polarity are tuned [8]. Needed to consider both specific and hazardous path of action of new released ILs upon the environment risk, see Appendix for an applied ecotoxicity review in general and of that of ILs in special, the SPECTRAL-SAR scheme of quantification of chemical–biological interaction is here suited to a specific IL cationic-anionic structural interaction.

SPECTRAL-SAR-IL METHODS

As previously was described, see [9] and the Chapters 6 and 7 of the present volume, when the ILs activity is evaluated two different additive models for modeling anionic-cationic interaction can be examined.

The Activity Additive |1+> Model

The first one is based on the vectorial summation of the produced anionic and cationic biological effects $|Y\rangle$, named the $|1+\rangle$ model, and which is constructed on the superposition of the anionic (subscripted with A) and cationic (subscripted with C) activities.

$$\left|Y_{AC}\right\rangle^{1+} = \left|Y_C\right\rangle + \left|Y_A\right\rangle \tag{8.1}$$

The analysis based on this model was reported for the *Vibrio fischeri* and *Daphnia magna* ecotoxicity [9].

The Structure Additive |0+> Model

The second SPECTRAL-SAR model, named $|0+\rangle$, is employed when the additive stage is considered at the examined Hansch factors $|X = LogP, POL, E_{TOT}\rangle$, which are first combined to produce the anionic-cationic (subscripted with AC) indices that are further used to produce the spectral mechanistic map of the concerned interaction [9 (b)]:

$$\left|Y_{AC}\right\rangle^{0+} = \hat{O}_{S-SAR} \left|0+\right\rangle = \hat{O}_{S-SAR} f\left(\left\{\left|X_A\right\rangle\right\}, \left\{\left|X_C\right\rangle\right\}\right) \tag{8.2}$$

with the particular specifications of the spectral vectors:

$$f\left(LogP_A, LogP_C\right) \equiv LogP_{AC} = \log\left(e^{LogP_A} + e^{LogP_C}\right) \in \left\{\left|X_{1AC}\right\rangle\right\}, \tag{8.3}$$

$$f\left(POL_A, POL_C\right) \equiv POL_{AC} = \left(POL_A^{1/3} + POL_C^{1/3}\right)^3 \in \left\{\left|X_{2AC}\right\rangle\right\} \text{ [Å}^3\text{]}, \tag{8.4}$$

$$f\left(E_A, E_C\right) \equiv E_{AC} = E_A + E_C - 627.71\frac{q_A q_C}{POL_{AC}^{1/3}} \in \left\{\left|X_{3AC}\right\rangle\right\} \text{ [kcal/mol]}. \tag{8.5}$$

The open issue addresses whether the $|0+\rangle$ and $|1+\rangle$ states yields with the same results or in which aspects they might differ in the IL ecotoxicity upon certain species. Nevertheless, a practically criteria of deciding upon activity or structure additivity models, modeled by equations (8.1) and (8.2), respectively, may be set respecting the so called *IL internal angle* between the anion-cationic activity vectors, with $y_{iA}, y_{iC}, i = \overline{1, N}$ components [9(a)]:

$$\cos\theta_{AC} = \frac{\sum_{i=1}^{N} y_{iC} y_{iA}}{\sqrt{\sum_{i=1}^{N} y_{iC}^2 \sum_{i=1}^{N} y_{iA}^2}} \begin{cases} \geq 0.707107 \ldots |0+\rangle \ MODEL \\ < 0.707107 \ldots |1+\rangle \ MODEL \end{cases} \tag{8.6}$$

for the N-molecules tested. The present study explores both these SPECTRAL-SAR approaches for modeling ILs-acetylcholinesterase interaction on *E. electricus* species.

APPLICATION ON ELECTRIC EEL

The SPECTRAL-SAR ecotoxicological principles [10] are applied and interpreted for the observed toxicity of the ILs of Table 8.1, with either $|0+>$ and $|1+>$ models. The experimental data regards the inhibition of acetylcholinesterase activity from the electric organ of the electric eel (*Electrophorus electricus*), Figure 8.1, as was measured using a colorimetric assay based on the reduction of 5,5A-dithio-bis-(2-nitrobenzoic acid) (DTNB) [3].

Figure 8.1. *Electrophorus electricus* from family of Gymnotidae (naked-back knifefishes) (photo taken at the Venezuela Pavilion at Expo2000 in Hannover by Heike Fьrderer); credit: http://www.fishbase.org/Summary/SpeciesSummary.php?id=4535.

The SPECTRAL-SAR of the ILs against of acetylcholinesterase activity (Table 8.1) and the associated computed spectral norms, the statistic and algebraic correlation factors [10], through for all possible correlation models are considered from the *cationic data, anionic data*, to model the cationic-anionic interaction for the $|1+>$ and $|0+>$ states, as previously described, respectively, with results displayed in Tables 8.1–8.6. Next, the spectral representation of the chemical–biological interaction paths across the SPECTRAL-SAR modeled endpoints is performed according with *least path rule* within the *spectral norm-correlation space* [10 (b), (c)], for the resulted IL $|1+>$ and $|0+>$ norms, for both the statistic and algebraic versions of correlation factors, see Table 8.7.

Table 8.1. The observed activity [3] and the Hansch structural parameters LogP, POL (E³), and E_{TOT} (kcal/mol) as accounting for the hydrophobicity, electronic (polarizability), and steric (total energy at optimized 3D geometry) effects [11], for each cation (subscripted as "C") and anion (subscripted as "A") fragment containing ionic liquids (subscripted as "AC") of Figure 8.2, as well as for the structure additive state, by means of equations (8.3–8.5), respectively.

No.	Full Name	Activity	LogP ($\lvert X_1\rangle$)			POL [Å3] ($\lvert X_2\rangle$)			$-E_{TOT}$ [kcal/mol] ($\lvert X_3\rangle$)		
		$\lvert Yexp\rangle$	$\lvert X_{1C}\rangle$	$\lvert X_{1A}\rangle$	$\lvert X_{1AC}\rangle$	$\lvert X_{2C}\rangle$	$\lvert X_{2A}\rangle$	$\lvert X_{2AC}\rangle$	$\lvert X_{3C}\rangle$	$\lvert X_{3A}\rangle$	$\lvert X_{3AC}\rangle$
1.	1-n-propyl-3-methylimidazolium tetrafluoroborate	2.28	0.28	1.37	1.65984	15.39	2.46	56.5053	236436.8594	261310.5938	497584
2.	1-n-butyl-3-methylimidazolium tetrafluoroborate	2.02	0.68	1.37	1.77652	17.22	2.46	60.8031	260646.6406	261310.5938	521798
3.	1-n-octyl-3-methylimidazolium tetrafluoroborate	1.66	2.26	1.37	2.60405	24.53	2.46	77.0645	357484.5938	261310.5938	618648
4.	1-n-decyl-3-methylimidazolium tetrafluoroborate	1.10	3.05	1.37	3.2209	28.23	2.46	84.883	405903.4375	261310.5938	667071
5.	1-n-benzyl-3-methylimidazolium tetrafluoroborate	1.97	1.25	1.37	2.00495	21.37	2.46	70.1876	330305.1875	261310.5938	591464
6.	1-phenyl-ethyl-3-methylimidazolium tetrafluoroborate	1.89	1.5	1.37	2.13026	23.21	2.46	74.2165	354514.625	261310.5938	615676
7.	1-n-butyl-3-ethylimidazolium tetrafluoroborate	2.03	1.02	1.37	1.90338	19.05	2.46	64.9974	284859.5313	261310.5938	546014
8.	1-n-butyl-3-methylimidazolium hexafluorophosphate	2.15	0.68	2.06	2.28441	17.22	1.78	54.623	260646.6406	580264.9375	840746
9.	1-n-butyl-3-methylimidazolium chloride	1.92	0.68	0.63	1.34846	17.22	2.32	59.5999	260646.6406	285190.781311	545677
10.	1-n-butyl-3-methylimidazolium bromide	1.90	0.68	0.94	1.51157	17.22	3.01	65.2647	260646.6406	1596918.25	1857410

Table 8.1. (Continued)

No.	Full Name	Activity $|Y_{exp}\rangle$	LogP ($|X_1\rangle$)			POL [Å3] ($|X_2\rangle$)			$-E_{TOT}$ [kcal/mol] ($|X_3\rangle$)		
			$|X_{1C}\rangle$	$|X_{1A}\rangle$	$|X_{1AC}\rangle$	$|X_{2C}\rangle$	$|X_{2A}\rangle$	$|X_{2AC}\rangle$	$|X_{3C}\rangle$	$|X_{3A}\rangle$	$|X_{3AC}\rangle$
11.	1-n-butyl-3-methylimidazolium octyl sulfate	1.98	0.68	4.23	4.25832	17.22	17.41	138.519	260646.6406	625818.4375	886344
12.	1-n-butyl-3-methylimidazolium trifluoromethyl sulfonate	1.94	0.68	3.05	3.13937	17.22	7.20	91.9356	260646.6406	1128283.625	1388790
13.	1-n-butyl-3-methylimidazolium dicyanoamide	1.95	0.68	0.43	1.25594	17.22	5.51	82.2317	260646.6406	147935.9844	408438
14.	1-n-butyl-3-methylimidazolium 2-(2-methoxyethoxy)ethyl sulfate	2.06	0.68	1.11	1.61108	17.22	13.18	120.881	260646.6406	645851.5	906371
15.	1-n-butyl-3-methylpyridinium tetrafluoroborate	1.53	3.32	1.37	3.45302	19.35	2.46	65.6761	274222.625	261310.5938	535378
16.	1-n-butyl-3-methylpyridinium hexafluorophosphate	1.45	3.32	2.06	3.56971	19.35	1.78	59.1642	274222.625	580264.9375	854326
17.	Trihexyl(tetradecyl)phosphonium dicyanoamide (cytec 105)	3.40	9.23	0.43	9.23015	60.27	5.51	183.926	987499.25	147935.9844	1135320
18.	Trihexyl(tetradecyl)phosphonium hexafluorophosphate (cytec 106)	3.30	9.23	2.06	9.23077	60.27	1.78	135.215	987499.25	580264.9375	1567640
19.	Trihexyl(tetradecyl)phosphonium tetrafluoroborate (cytec 111)	3.47	9.23	1.37	9.23039	60.27	2.46	146.418	987499.25	261310.5938	1248690

Table 8.2. Spectral structure-activity relationships (SPECTRAL-SAR) of the ionic liquids against of acetylcholinesterase activity from the electric organ of the electric eel (*Electrophorus electricus)* and the associated computed spectral norms with statistic and algebraic correlation factors [10 (c), (d)] through all possible correlation models considered from the *cationic data* in Table 8.1.

Mode	Vectors[*]	Cationic SPECTRAL-SAR	$\|\|Y_C>^{MODE}\|\|$	$r_{S-SAR}^{STATISTIC}$	$r_{S-SAR}^{ALGEBRAIC}$ [*]
Ia	$\|X_0>$, $\|X_{1C}>$	$\|Y_C>^{Ia} = 1.71403\|X_0>$ $+0.151303\|X_{1C}>$	9.3907	0.743456	0.982248
Ib	$\|X_0>$, $\|X_{2C}>$	$\|Y_C>^{Ib} = 1.25198\|X_0>$ $+0.0331507\|X_{2C}>$	9.43784	0.822285	0.987178
Ic	$\|X_0>$, $\|X_{3C}>$	$\|Y_C>^{Ic} = 1.30338\|X_0>$ $-2.01381\cdot10^{-6}\|X_{3C}>$	9.45334	0.846693	0.9888
IIa	$\|X_0>$, $\|X_{1C}>,\|X_{2C}>$	$\|Y_C>^{IIa} = 0.723168\|X_0>$ $-0.22236\|X_{1C}>+0.0760339\|X_{2C}>$	9.46156	0.859364	0.98966
IIb	$\|X_0>$, $\|X_{1C}>,\|X_{3C}>$	$\|Y_C>^{IIb} = 0.850384\|X_0>$ $-0.236646\|X_{1C}>-4.68839\cdot10^{-6}\|X_{3C}>$	9.48661	0.89694	0.99228
IIc	$\|X_0>$, $\|X_{2C}>,\|X_{3C}>$	$\|Y_C>^{IIc} = 2.30273\|X_0>$ $-0.439643\|X_{2C}>-2.79231\cdot10^{-5}\|X_{3C}>$	9.5503	0.986459	0.998941
III	$\|X_0>$, $\|X_{1C}>,\|X_{2C}>$, $\|X_{3C}>$	$\|Y_C>^{III} = 2.4248\|X_0>$ $+0.0350203\|X_{1C}>-0.46385\|X_{2C}>$ $-2.89539\cdot10^{-5}\|X_{3C}>$	9.55073	0.987037	0.998986

[*] $\|X_0>=\|1\ 1\ ...\ 1_{19}>$; [*]computed upon formula $\|\ \|Y_C>^{MODE}\ \|\|/\|\ \|Y_{EXP}>\ \|\|$ with $\|\ \|Y_{EXP}>\ \|\|=9.56042$.

Table 8.3. The same set of SPECTRAL-SAR information as in Table 8.2, here for the *anionic data* in Table 8.1.

Mode	Vectors[*]	Anionic SPECTRAL-SAR	$\|\|Y_C>^{MODE}\|\|$	$r_{S-SAR}^{STATISTIC}$	$r_{S-SAR}^{ALGEBRAIC}$ [*]
Ia	$\|X_0>$, $\|X_{1A}>$	$\|Y_A>^{Ia} = 2.21943\|X_0>$ $-0.0739556\|X_{1A}>$	9.18093	0.104743	0.960306
Ib	$\|X_0>$, $\|X_{2A}>$	$\|Y_A>^{Ib} = 2.1073\|X_0>$ $-4.73181\cdot10^{-4}\|X_{2A}>$	9.17663	0.00314251	0.959857
Ic	$\|X_0>$, $\|X_{3A}>$	$\|Y_A>^{Ic} = 2.18576\|X_0>$ $+1.76401\cdot10^{-7}\|X_{3A}>$	9.18074	0.102437	0.960286
IIa	$\|X_0>$, $\|X_{1A}>,\|X_{2A}>$	$\|Y_A>^{IIa} = 2.21333\|X_0>$ $-0.0986301\|X_{1A}>+0.0102865\|X_{2A}>$	9.18228	0.12007	0.960447
IIb	$\|X_0>$, $\|X_{1A}>,\|X_{3A}>$	$\|Y_A>^{IIb} = 2.25033\|X_0>$ $-0.0556466\|X_{1A}>+1.29662\cdot10^{-7}\|X_{3A}>$	9.18288	0.126365	0.960511
IIc	$\|X_0>$, $\|X_{2A}>,\|X_{3A}>$	$\|Y_A>^{IIc} = 2.17587\|X_0>$ $+0.00317468\|X_{2A}>+1.84606\cdot10^{-7}\|X_{3A}>$	9.18091	0.104476	0.9603004
III	$\|X_0>$, $\|X_{1A}>,\|X_{2A}>$, $\|X_{3A}>$	$\|Y_A>^{III} = 2.24547\|X_0>$ $-0.081245\|X_{1A}>+0.0110938\|X_{2A}>$ $+1.36833\cdot10^{-7}\|X_{3A}>$	9.18445	0.141282	0.960674

[*] $\|X_0>=\|1\ 1\ ...\ 1_{19}>$; [*]computed upon formula $\|\ \|Y_C>^{MODE}\ \|\|/\|\ \|Y_{EXP}>\ \|\|$ with $\|\ \|Y_{EXP}>\ \|\|=9.56042$.

Table 8.4. Spectral structure-activity relationship (SPECTRAL-SAR) of the ionic liquids for all possible correlation models considered for the cationic-anionic interaction within the activity additive |1+> state employing the results of Tables 8.2 and 8.3 and the equation (8.1).

Mode	Ionic Liquid |1+> SPECTRAL-SAR	$\| \| Y_{AC}>^{MODE} \| \|$	$r_{S-SAR}^{STATISTIC}$	$r_{S-SAR}^{ALGEBRAIC}$ *
Ia	$\|Y_{AC}>^{Ia} = \|Y_A>^{Ia} + \|Y_C>^{Ia}$	18.4674	3.34175 i	1.93165
Ib	$\|Y_{AC}>^{Ib} = \|Y_A>^{Ib} + \|Y_C>^{Ib}$	18.4854	3.32191 i	1.93354
Ic	$\|Y_{AC}>^{Ic} = \|Y_A>^{Ic} + \|Y_C>^{Ic}$	18.502	3.31937 i	1.93528
IIa	$\|Y_{AC}>^{IIa} = \|Y_A>^{IIa} + \|Y_C>^{IIa}$	18.5051	3.31406 i	1.9356
IIb	$\|Y_{AC}>^{IIb} = \|Y_A>^{IIb} + \|Y_C>^{IIb}$	18.5217	3.30655 i	1.93733
IIc	$\|Y_{AC}>^{IIc} = \|Y_A>^{IIc} + \|Y_C>^{IIc}$	18.5476	3.27711 i	1.94004
III	$\|Y_{AC}>^{III} = \|Y_A>^{III} + \|Y_C>^{III}$	18.5533	3.27854 i	1.94064

*computed upon formula $\| \| Y>^{Mode} \| \| / \| \| Y_{EXP}> \| \|$ with $\| \| Y_{EXP}> \| \| = 9.56042$.

Table 8.5. The values of the integral ionic liquid cosines angle, according with the equation (8.6), for all considered modes of action of the cationic and anionic activities from Tables 8.2 and 8.3, for the |1+> states of the ionic liquids of Figure 8.1.

Mode	Ia	Ib	Ic	IIa	IIb	IIc	III
$cos\theta_{AC}$	0.977604	0.972358	0.971751	0.970344	0.968455	0.960961	0.961348

Table 8.6. Spectral structure-activity relationships (SPECTRAL-SAR) of the ionic liquids for all possible correlation models considered for the cationic-anionic interaction within the structure additive |0+> model of equation (8.2) employing the structure data of Table 8.1 on equations (8.3–8.5).

Mode	Ionic Liquid |0+> SPECTRAL-SAR	$\| \| Y_{AC}>^{MODE} \| \|$	$r_{S-SAR}^{STATISTIC}$	$r_{S-SAR}^{ALGEBRAIC}$ *
Ia	$\|Y_{AC-Ia}>^{0+} = 1.469\|X_0> + 0.184781\|X_{1AC}>$	9.4185	0.790857	0.985156
Ib	$\|Y_{ACIb}>^{0+} = 1.02654\|X_0> + 0.0121125\|X_{2A}>$	9.37567	0.716589	0.980675
Ic	$\|Y_{AC-Ic}>^{0+} = 1.49483\|X_0> - 7.14467 \cdot 10^{-7}\|X_{3AC}>$	9.25926	0.460272	0.9685
IIa	$\|Y_{AC-IIa}>^{0+} = 1.28923\|X_0> + 0.144321\|X_{1AC}> + 0.00358292\|X_{2AC}>$	9.42421	0.80025	0.985753
IIb	$\|Y_{AC-IIb}>^{0+} = 1.39785\|X_0> + 0.175543\|X_{1AC}> - 1.2051 \cdot 10^{-7}\|X_{3AC}>$	9.42021	0.793675	0.985334
IIc	$\|Y_{AC-IIc}>^{0+} = 0.91542\|X_0> + 0.0108015\|X_{2AC}> - 2.66714 \cdot 10^{-7}\|X_{3AC}>$	9.38468	0.732808	0.981618
III	$\|Y_{AC-III}>^{0+} = 1.23314\|X_0> + 0.137621\|X_{1AC}> + 0.00346107\|X_{2AC}> -1.05349 \cdot 10^{-7}\|X_{3AC}>$	9.4255	0.802368	0.985888

*computed upon formula $\| \| Y>^{Mode} \| \| / \| \| Y_{EXP}> \| \|$ with $\| \| Y_{EXP}> \| \| = 9.56042$.

Table 8.7. Synopsis of the statistic and algebraic values of paths connecting the SPECTRAL-SAR models of Tables 8.4 and 8.6 in the norm-correlation spectral-space for the ionic liquids against of acetylcholinesterase activity from the electric organ of the electric eel (*Electrophorus electricus*) throughout all the possible correlation models considered from the *anionic, cationic,* and *ionic liquid* additive *activity* $|1+>$ and *structure* $|0+>$ models, respectively; the *primary, secondary,* and *tertiary*—the so called alpha (α), beta (β), and gamma (γ)—paths are indicated accordingly with the least principle in spectral norm-correlation space for the statistic and algebraic variants of the correlation factors used, respectively.

Path	Cationic		Anionic		Ionic Liquid			
					state $|0+>$		state $|1+>$	
	statistic	algebraic	statistic	algebraic	statistic	algebraic	statistic	algebraic
Ia-IIa-III	0.291581	0.160901	0.036709 α	0.00353909 α	0.0134705	0.00703621	0.0582483 γ	0.086427
Ia-IIb-III	0.29157 γ	0.160901	0.036709	0.00353909	0.0134705 α	0.00703621 α	0.0560458	0.086427
Ia-IIc-III	0.291448	0.160901 γ	0.0372431	0.00358302	0.147841	0.0750578	0.0530531	0.086427 γ
Ib-IIa-III	0.199745 β	0.113504	0.138386	0.00785625	0.0992051 β	0.0501068	0.0506435	0.0682757
Ib-IIb-III	0.199756	0.113504 β	0.138379 γ	0.00785625 γ	0.0992069	0.0501068	0.0475165 β	0.0682757 β
Ib-IIc-III	0.199718	0.113504	0.138399	0.00785625	0.0992086	0.0501068 β	0.0486454	0.0682757
Ic-IIa-III	0.170829	0.0979132 α	0.0390218	0.00372716	0.38036	0.167146 γ	0.0325599+R* i	0.0515569 α
Ic-IIb-III	0.170843	0.0979132	0.0390218	0.00372716	0.380395	0.167146	0.0295517	0.0515569
Ic-IIc-III	0.170821 α	0.0979132	0.0390209 β	0.00372716 β	0.380663 γ	0.167146	0.022493 α	0.0515569

*R = 0.00431574.

Note the statistical correlation factors always yield lower values than the corresponding algebraically ones [10 (d)]. However, with the $|1+>$ model there are even record imaginary statistical correlation factors (in all the cases, all the modes), whereas the algebraically correlation outputs give supraunitary values due to the sum of the anionic and cationic norm lengths length, that is by doubling the intensity of the computed endpoint activity (Table 8.4). This can be phenomenological explained by the so called *resonance effect* when almost zero angle between the anionic and cationic endpoint vectors, for all molecular modes of actions of Tables 8.2 and 8.3 are recorded as clearly evidenced by the cosines values of equation (8.6) [9(a)], see Table 8.5. This is giving the sign that the $|0+>$ model should be the proper one here modeling the ecotoxicological effects, see Table 8.6; similar phenomenological results were previously reported on *Daphnia magna case*, too [9(b)].

Worth remarking that the $|0+>$ and $|1+>$—SAR models furnish different hierarchies of the paths for the chemical–biological actions, see Table 8.7, being coincident only for the beta path (in comparison, for instance with the *Daphnia* case we obtained the same results, however with different set of ILs compounds [9(b)]). As well, the statistical imaginary correlation values for the ILs $|1+>$ states in Table 8.4 extend their behavior in the Table 8.7 too, but only in one case, namely *Ic-IIa-III*; nevertheless, that path is excluded from the validation principle (minimum path search) [9(b)].

Overall, for the IL $|1+>$ state the cationic influence is found with the dominant contribution over the anionic effects in ecotoxicity through the ecotoxicological paths in cationic and anionic subsystems are summed up in the paths of the corresponding ILs in a not trivial ways:

$$\alpha_C + \beta_A = \alpha_{AC}^{|1+\rangle}, \beta_C + \gamma_A = \beta_{AC}^{|1+\rangle}, \gamma_C + \alpha_A = \gamma_{AC}^{|1+\rangle}. \tag{8.7}$$

Actually, ecotoxicologically, the anionic-cationic interaction paths containing $|1+>$ state of ILs can be interpreted as: (i) the anionic paths effect is marginal over the cationic paths, in all studied cases; (ii) the IL alpha path originates with the steric (cationic) arrangements, through the total energy at the optimum energy, followed by the beta path rooting on the (cationic) long range interaction or ionic interaction or van der Waals interaction, through the degree of polarizability, ending with the gamma path quantifying the cell membrane by the (cationic) hydrophobicity. Here, only the beta path equally behaves within statistic and algebraic description for all intermediate endpoints.

Instead, for the IL $|0+>$ state the anionic influence is found as the main contributor over the cationic effects in combined ecotoxicity of the alpha path,

$$\gamma_C + \alpha_A = \alpha_{AC}^{|0+\rangle}, \beta_C + \gamma_A = \beta_{AC}^{|0+\rangle}, \alpha_C + \beta_A = \gamma_{AC}^{|0+\rangle}, \tag{8.8}$$

while the cationic effect is dominant over anion for beta path, and canceling with anion on gamma path. Moreover, the IL $|0+>$ alpha path originates with the (anionic) cell membrane passport quantified by the hydrophobicity, followed by the (cationic) polarizability influence on beta path, collapsing on the gamma path as (IL) steric arrangements through the anionic-cationic (canceling) interaction. Now, only the alpha path displays full identical statistic and algebraic description for all intermediate endpoints.

Finally, the Figure 8.3 (a–c) collects all SPECTRAL-SAR information in either $|1+>$, $|0+>$, and combined (viewed as excited-ground state) states, respectively, in terms of computed norms for all considered endpoints, the observed (experimental) one included. Note that the hypersurface in Figure 8.3(c) was generated observing that (in average) the IL $|0+>$ state norms are placed between those of anionic and cationic ones, far by the computed norms of ILs of Figure 8.2 in $|1+>$ state. In this respect the alpha, beta, and gamma paths on hypersurface of Figure 8.3(c) are those obtained from algebraic state $|0+>$ of Table 8.7 being assumed as those providing the complete norm-based ecotoxicological mechanisms.

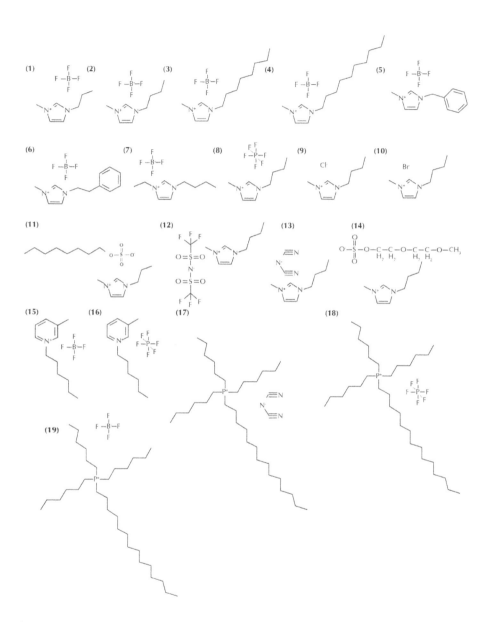

Figure 8.2. The series of the ionic liquids of that toxic activities A = Log(EC$_{50}$) on acetylcholinesterase activity (EC 3.1.1.7) from the electric organ of the electric eel (*Electrophorus electricus*) were considered, see Table 8.1 [3].

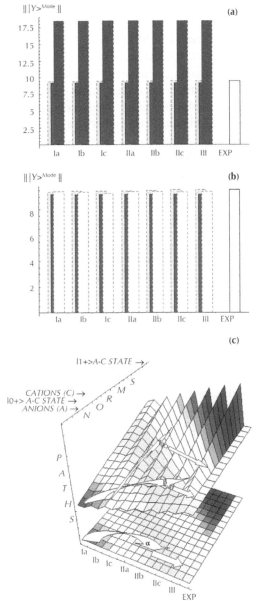

Figure 8.3. (a) The norm bar representations of the cationic (on first place), anionic (on second place) for the |1+> state of the ionic liquids (on third place) of Figure 8.1, for computed and experimental endpoints of Tables 8.2 and 8.3; (b) repeated picture of (a) for ionic liquids of Figure 8.2 within the |0+> state and the norms of Table 8.6; (c) the hypersurface of anionic, cationic, and ionic liquid |0+> and |1+> states with the alpha, beta, and gamma paths drawn for the algebraic |0+> state hierarchically of Table 8.7.

CONCLUSION

In the context of QSAR ecotoxicological studies the SPECTRAL-SAR approach is engaged to produce the $|0+>$ and $|1+>$ SPECTRAL-SAR-IL states and to apply them on the particular case of electric eel species.

While being different of cationic-anionic additive structure $|0+>$ state the additive activity $|1+>$ model gives reliable results when the so called internal anionic-cationic angle is exceeding $45°$ and producing the so called resonance effect (imaginary statistical factors, supra-unitary algebraic factors) otherwise.

The present ecotoxicological analysis reveals the primary alpha, beta, and gamma paths as rooting on hydrophobicity (cell' trans-membrane transporting), polarizability (electrostatic interaction), and steric (optimum configuration) effects, respectively, within the algebraic norm-based structure additive state $|0+>$.

APPENDIX: SURVEY ON ECOTOXICITY AND GREEN CHEMISTRY

Ecotoxicology is the science of the impact of toxic substances on living organisms, encompassing all levels of biological organization from single organisms to ecosystems. Ecotoxicology integrates environmental chemistry, biochemistry, toxicology, and ecology in a multidisciplinary manner. The unifying theme of ecotoxicological research is to provide general concepts to evaluate the potential harmfulness of pollutants. Traditionally ecological risk assessment consists of three steps:

  i. Hazard identification including the collection and evaluation of all available information on the given chemical to assess its potential adverse effects;
  ii. Exposure and effect assessment, which results in a predicted environmental concentration (PEC) and predicted no-effect concentration (PNEC); and
  iii. Comparison of PEC and PNEC to characterize the risk imposed by a given chemical [12].

In the case of new chemicals, we have to test their toxicity in all the surroundings that they could reach a destination: soil, fresh, and salt water. Knowing that not all-relevant target sites are found in a single organism, it is widely recognized that no single bioassay can be used to assess the toxic effects of every different mode of action. Therefore a battery of test methods is probably required [12, 13].

Comparing the concentration in an environmental compartment with a predicted no-effect concentration we are performing the risk assessment of chemicals. The latter value is generally obtained by using results from laboratory tests with species representative for that environmental compartment and applying to them an assessment factor, to account for the uncertainties involved in extrapolating from mono-species laboratory tests to multi-species ecosystems. The corresponding directives and guidance documents of the EU (Technical Guidance Document (TGD) for new and existing chemicals and biocides and the Plant Protection Products Directive (91/414/EEC) for pesticides) describes how exposure and effect concentrations should be derived. General adverse effects on the terrestrial environment include, according to the Scientific Committee on Toxicology, Ecotoxicology, and the Environment [15]:

- Effects on soil functions, and particularly on the capacity of soil to act as substrate for plants including effects on seed germination, and those on organisms (invertebrates, micro-organisms) important for proper soil function and nutrient cycle conservation,
- Effects on plant biomass production, related to contamination of soil or air including deposition on plant surfaces,
- Effects on soil, above-ground, and foliar invertebrates, which represent food for other organisms, and cover essential roles as pollinators, detrivores, saprophages, pest controllers, and so forth,
- Effects on terrestrial vertebrates exposed to contaminated food, soil, air, water, or surfaces, with obvious economic and/or social consequences, and
- Accumulation of toxic compounds in food items and through the food chain.

Although the terrestrial ecosystem is composed of an above-ground soil and a groundwater community, the TGD only addresses risk to soil organisms. For most chemicals, data on toxicity to soil organisms is limited. *Toxicity tests with plants and earthworms can be requested for new substances at a production volume of 100 t/a.* However, up to now, only a relatively small number of new chemicals have reached that level. For existing chemicals, the database is often even poorer. An evaluation of the first EU priority list has found that two or more tests with soil organisms, as needed to derive a predicted no effect concentration (PNEC), are available for only about 35% of these high priority, high production volume chemicals [16]. Therefore, in the absence of test data with soil organisms, the TGD suggests use of the equilibrium partitioning method to extrapolate soil toxicity from the aquatic toxicity [16, 17].

Natural soil is one of the key elements enabling life on earth. It plays a central role in all terrestrial systems, functions as a habitat for micro-organisms, plants and animals, and as a water filter and store. Many sites are severely polluted by heavy metals, pesticides, PAH, PCBs, explosives, and surfactants, especially those soils influenced by industry, traffic, and domestic heating. Contamination increases in topsoil in the following order:

arable soils < mineral soil below forest < permanent grassland < urban soils < specifically contaminated sites.

Risk assessment of field soil, contaminated soil, and soil material obtained by remediation techniques, cannot be based on the chemical-analytical determination of the total contents of specific contaminants, because they do not provide information of either unexpected substances, among others of metabolites of pollutants, and of their bioavailable fractions.

Therefore, the chemical-analytical determinations of the total pollutant content are not sufficient for the qualitative and quantitative detection of potential risks. Risk assessment has to consider the mobility and bioavailability of the contaminants, and their uptake and metabolization by micro-organisms, plants, and animals including man. For an appropriate consideration of all these factors, biotests have to be applied which reflect the pollutant-induced reduction of the habitat function [18] and the retention function [19] of soil or soil material. To consider both functions, a battery of evaluated biotests is obligatory, which integrates all adverse effects caused by contaminated soil [20]. For the time being, a minimum ecotoxicological test battery should at least include bacteria, vegetables, invertebrates, and mammalian and not-mammalian cells [14]. *Examples of development of tests batteries for risk assessment of soil and soil material from literature follow bellow.*

The implementation of the Bundes-Bodenschutzgesetz in Germany made it necessary to establish projects for the development of a test battery for the assessment of soil quality. This battery is based on existing biotests for the assessment of chemical substances, which had to be modified for field soils, and new biotests, which had to be developed. Three different projects for battery development were established between 1996 and 2001, and the results were published mostly in German. More than 20 biotests were modified developed and evaluated in these projects. The retention function of soil material was assessed with soil eluates by aquatic tests and tests for genotoxicity. The habitat function was analyzed by tests with micro-organisms, terrestrial invertebrates, and plants. In addition, reproduction tests of short duration with nematodes, screening tests, and a DNA array assay (Celegans ToxChip) were developed [20]. A battery of test systems and indicators would be representative of a wide range of organisms.

Another example of an ecotoxicological test battery was finding by Repetto G. et al. in 2001 [14]. The battery included the immobilization of *Daphnia magna*, bioluminescence inhibition in the bacterium *Vibrio fischeri*, growth inhibition of the alga *Chlorella vulgaris*, and micronuclei induction and root growth inhibition in the plant *Allium cepa*. Cell morphology, total protein content, MTS metabolization, lactate dehydrogenase leakage and activity, and glucose-6-phosphate dehydrogenase activity were studied in the salmonid fish cell line RTG-2. The total protein content, LDH activity, and MTT metabolization in Vero monkey kidney cells were also investigated.

First of all, the test systems were used for ecotoxicological evaluation of pentachlorophenol. The system most sensitive to pentachlorophenol was micronuclei induction in *A. cepa*, followed by *D. magna* immobilization, and bioluminescence inhibition in *V. fischeri* bacteria [14]. Second, carbamazepine, an anticonvulsant commonly present in groundwater, was studied, using the same six ecotoxicological model systems with the 18 endpoints evaluated at different exposure time periods. The most sensitive system to carbamazepine was the Vero cell line, followed by *Chlorella vulgaris, Vibrio fischeri, Daphnia magna, Allium cepa*, and RTG-2 cells [21]. Also, in 2005, these authors used this battery for ecotoxicological evaluation of additive butylated hydroxyanisole. The most sensitive system to this additive was the inhibition of bioluminescence in *Vibrio fischeri* bacteria, which resulted in an acute low observed adverse effect concentration (LOAEC) of 0.28 mM. The next most sensitive system was the immobilization of the cladoceran *Daphnia magna* followed by: the inhibition of the growth of the unicellular alga *Chlorella vulgaris*. The differences in sensitivity for the diverse systems that were used (EC50 ranged from 1.2 to more than 500 mM) [22]. So, the same test battery was used for the ecotoxicologically evaluation of different type of substances.

Assessment of potential impacts in the receiving water to which an effluent is discharged requires inclusion in the test battery of species representing different trophic levels: typically algae, which are primary producers, invertebrates, primary consumers and fishes, secondary consumers. The assessment of the potential effects of effluents on different trophic levels was agreed at the direct toxicity assessment (DTA) Demonstration Program (Environment Agency 1999) being settled with the approach to hazard assessment used in the management and regulation of new and existing chemicals.

Regulators from Europe recognize the use of the different taxonomic groups, Umweltbundesamt 1997, as well as Canada, Environment Canada 1999 and United State, USEPA 1991. But also, the way that bioassay are used (test design and quality assurance/quality control applied) needs to be tailored for different applications. The test methods used for ecotoxicological test batteries are: for freshwater, 72 h algae (*Pseudokirchneriella subcapitata)* growth inhibition test (OECD 201), 42 h *Daphnia magna* immobilization (OECD 202), and 96 h Juvenile rainbow trout (*Onchorhynchus mykiss*) lethality (OECD 203); in the case of saltwater (estuarine/marine) 72 h algae (*Skeletonema costatum*) growth inhibition test (ISO 10253), 24 h Oyster (*Crassostrea gigas*) embryo-larval development (ICES, Vol. 11), and 94 h Juvenile turbot (*Scophthalmus maximus*) lethality (ISO/WD 15990). These methods are widely used in regulatory chemical hazard assessment schemes and all have internationally recognized and standardized protocols that have been promulgated by bodies such as the Organisation for Economic Co-operation and Development (OECD), relate to the evaluation of pure substances, International Standard Organization (ISO) and the International Council for the Exploration of the Sea (ICES), procedures that can be applied to pure substances as well to environmental samples [13]. Evaluation of the adverse effects of pollutants in aquatic ecosystems requires discrimination between the total concentration of a chemical, the bioavailable fraction, and the final concentration at the target site(s) [23].

So, we can choose from these presented methods for integrated them in a future ecotoxicological test battery. Until now, we have presented examples of chosen ecotoxicologically test batteries for hazard assessment of new chemicals. Recently, another very interesting strategy for ecotoxicology evaluation, namely the test-kit-concept was advanced where a major role is dedicated to the test battery ecotoxicity/toxicity tool, which has to allow a systematic evaluation of (eco) toxicological properties of chemicals on biological systems of different complexity [7]. The studies with test battery have to be performed at different levels: molecular, cellular, organism, and ecosystem. Although the test-kit-concept is an iterative process, one of the procedures that have to be performed from the very beginning is the test of the effect of the target chemical at molecular level. This can be realized studying the effect of that compound on some well choosed enzymes. To make a good enzymes kit for ecotoxicological test battery, the selected enzymes has to fulfill some requirements:

- to be part from a major biochemical pathway,
- to be found in almost all organisms, and
- to be extracted/purified from a cheep biological source.

Multienzymatic test battery as a model for testing at molecular level the toxicity of chemicals compounds was also initiated to increase the certainty with which the new compounds affects the environment (the living organisms), while being necessarily to work on a small number of enzymes but those that are part of major biochemical pathways and are found in almost all organisms [24, 25]. Liver alkaline phosphatase, spinach catalase, alcohol dehydrogenate from yeast, and porcine trypsin seems to fulfill the entire requirements for a good candidate for multienzymatic test battery kit. For testing the adverse effects of new compounds on environment, it will be easy (and cheaper) to start with the study of the effects at molecular level of these chemicals on some enzyme activities, selected by definite criteria that will allow the reductions of experiments.

On the other side, taking in consideration that the toxicity of the new chemicals depends of the calculated inhibitions constants, the way in which the new kinetics parameters associated with respectively enzymatic reaction are calculated, it is critical for a good practically prediction. The application of the quasi-steady-state approximation (QSSA) in biochemical kinetics allows the reduction of a complex biochemical system with an initial fast transient into a simpler system. The simplified system yields insights into the behavior of the biochemical reaction, and analytical approximations can be obtained to determine its kinetic parameters. However, this process can lead to inaccuracies due to the inappropriate application of the QSSA [26]. The main problem with the Michaelis–Menten equation is that it accounts only for the velocity of the initial time of the reaction. The information that is outside the first moments of the progress curve is virtually lost or neglected. More, such way, when velocities are measured it is usually to determine one velocity from each experimental assay. In special in the case of sensitive enzymes, this procedure is difficult to be performed. Another complication of the Michaelis–Menten equation is that, even describing a kinetic, differs than the ordinary chemical ones by its rectangular hyperbola shape, rather by an expected exponential form. In the context of Michaelis–Menten mechanisms a new related enzymatic kinetic is proposed; see Appendix of the Chapter 3 of the present volume. It provides an exponential form of the instantaneous velocity for a reaction counting for the competitive and uncompetitive inhibitions of the enzyme and enzyme-substrate complex, respectively, as well for the reversible overall process. Such an approach is suited for progress curves analysis of the substrate and product concentrations [27, 28].

Also within the context of the Michaelis–Menten mechanism the reversible enzymatic kinetics is studied to provide the close form solution under the W-Lambert function, as well as the equivalent new proposed logistic, see Appendix of the Chapter 3 of the present volume. It produces useful tool with which the characteristic parameter estimation can be made in an accurate straightforward manner so decreasing the number of associated experimental assays. Such approach serves as a proper analytical framework for describing of many important biochemical pathways, for example the regulation of carbohydrate metabolism, as any regenerative cyclical and necessarily reversible-natural process. So, in the future, worth that toxicity be evaluated at the enzymatic level, by calculation of the inhibition constants, with the new logistical transformation.

Finally, going to study the new substances' ecotoxicity as the ILs display since they, as new generations of solvents have the capability to replace organic solvents in biochemical synthesis and catalysis [29]. ILs are composed by a cationic structure (imidazolium, pyridinium, pyrolidinium, phosphonium, and ammonium) with alkyl or arylalkyl hydrofobic substituents, and an anionic structure, inorganic, or organic (see Figure 8.A1).

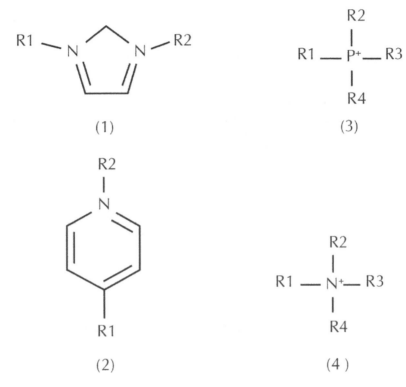

Figure 8.A1. The four important classes of ionic liquids, function of the cationic structure: (1) imidazolium, (2) pyridinium, (3) phosphonium, (4) ammonium.

Due to definition all organic salts with a melting point bellow 100°C belongs to this group of chemicals. An important convenience of ILs is their negligible vapor pressure, which decreases the risk of technological exposure and of solvent loss to the atmosphere [30]. These advantages make ILs associated with the term "green", and give opportunity for the elimination of volatile organic compounds (VOCs) and a future production at industrial scale. As such, *green chemistry* is the effort of reducing or eliminating the use or generation of hazardous substances in the design, manufacture, and application of chemical products [31]. But until now the interaction with the environment is not very well yet tested, so this is an open field for ecotoxicological scientists. For completion, few properties of ILs are reviewed here [8]:

- They present under combination of large organic positive ions—such as for example 1-octyl-3-methyl imidazolium [OMIM]⁻ and smaller inorganic or organic negative ions, like for example tetrafluoroborate (BF4⁻);

- They are non-corrosive, and are characterized by low viscosity; and
- They dissolve a wide range of organic molecules to an appreciable extent, meaning much lower volumes of solvent are required for a given process.

Thus, the future of ILs is that for an unpolluted technology is overwhelmed with reducing the waste from an industrial chemical process to minimal. Its implementation will lead to a cleaner environment. The sustainable development should give improvements to the economical, ecologically, and socially conditions, for the present and future generations. That will be the chance to take into consideration the technology and in the same time will be avoided the hazardous structures from the very beginning of development of industrial chemical products. Together the chemists, biologists, and technologists must find goon-fitting compounds on the base of SAR and QSAR algorithm results. The structures of those compounds will be compared and the properties will be theoretically presumed [7]. The (eco) toxicological tests of IL entities aim for the identification of the individual effects of different cations (headgroups, with identical side chains and anions), anions (with identical cations), and systematic variations of R1, R2, and so forth at identical head groups and with identical anions. In 2004 we published our first systematic investigations of the influence of an increasing chain length of R1 or R2 of the imidazolium cation moiety on the cytotoxicity in marine bacteria and two types of mammalian cell cultures [32]. Was became that the shorter the chain length(s) of side chain(s), the lower the cytotoxicity. Following the concept of a flexible test battery was also investigated the side chain length effect at the molecular level, selecting acetylcholine esterase as a model enzyme. This enzyme was chosen because of the similarity of the chemical structure of many IL cations and acetylcholine. Again the same results were obtained. Jastorff and his collaborators were done different ecotoxicologically studies on ILs [3, 7, 33, 34], but it is still stays as an open field due the multitude possible chemical combinations for the constitution of different ILs.

KEYWORDS

- **Anionic-cationic interaction**
- **Ecotoxicology**
- **Electric eel**
- **Ionic liquid modeling**

PERMISSIONS

This Chapter was previously published in International Journal of Chemical Modeling, 2(1) (2010) 85-96 as Spectral-SAR Ecotoxicology of Ionic Liquids-Acetylcholine Interaction on E. Electricus Species.

PART III
QUANTUM AND STATISTICAL INTERPRETATION OF SPECTRAL-SAR METHOD

Chapter 9

From SPECTRAL-SAR to QUANTUM-SAR Algorithm: Designing the Polyphenolic Anticancer Bioactivity

Mihai V. Putz, Ana-Maria Putz, Marius Lazea, Luciana Ienciu, and Adrian Chiriac

INTRODUCTION

Aiming to assess the role of individual molecular structures in the molecular mechanism of ligand-receptor interaction correlation analysis, the recent SPECTRAL-SAR approach is employed to introduce the QUANTUM-SAR (QUAntum Nature of TUning Metabolism by SAR or even simpler as QuaSAR) "wave" and "conversion factor" in terms of difference between inter-endpoint inter-molecular activities for a given set of compounds; this may account for inter-conversion (metabolization) of molecular (concentration) effects while indicating the structural (quantum) based influential/detrimental role on bio-/eco-effect in a causal manner rather than by simple inspection of measured values; the introduced QuaSAR method is then illustrated for a study of the activity of a series of flavonoids on breast cancer resistance protein (BCRP).Being used in Chemistry during the second half of twentieth century as an extended statistical analysis [1–8], the quantitative structure-activity relationship (QSAR) method had attained in recent years a special status, officially certified by European Union as the main computational tool (within the so called "*in silico*" approach) for the regulatory assessments of chemicals by means of non-testing methods [9–15].

However, while QSAR primarily uses the multiple regression analysis [6–8], alternative approaches as such neuronal-network (NN) or genetic algorithms (GA) have been advanced to somehow generalize the QSAR performance in delivering a classification of variables used, in the sense of principal component analysis (PCA) and partial least squares (PLS) methodologies; still, the claimed advantage of the NN over QSAR techniques is limited by the fact that the grounding physical–mathematical philosophies are different since highly non-linear with basic multi-linear pictures are compared, respectively [16–23]. Actually, the chemical–physical advantage of QSAR stands in its multi-linearity correlation that resembles with superposition principle of quantum mechanics, which allow meaningful interpretation of the structural (inherently quantum) causes associated with the latent or unobserved variables (sometimes called as *common factors*) into the observed effects (activity) usually measured in

Putz M.V., Putz A.M., Lazea M., Ienciu L., Chiriac A. "Quantum-SAR Extension of the Spectral-SAR Algorithm. Application to Polyphenolic Anticancer Bioactivity", *International Journal of Molecular Sciences*, 10(3) (2009) 1193-1214; DOI: 10.3390/ijms10031193.

terms of 50%-effect concentration (EC_{50}), associated with various types of bioaccumulation and toxicity [24].

Nevertheless, many efforts have been focused on applying QSAR methods to nonlinearity features from where the "expert systems" emerged as formalized computer-based environments, involving knowledge-based, rule-based, or hybrid automata able to provide rational predictions about properties of biological activity of chemicals or of their fragments; it results in various QSAR based databases: the model database (QMDB)—inventorying the robust summaries of QSARs that can be appealed by envisaged endpoint or chemical, the prediction database (QPDB)—when data from QMDB are used for further prediction to be stored, or together towering the chemical category database (CCD) documentation [25–31].

Therefore, although undoubtedly useful, the "official" trend in employing QSAR methods is to classify, over-classify and validate through (external or molecular test set) prediction, a gap between the molecular computed orderings and the associate mechanistic role in bio-/eco-activity assessment remains as large as the QSAR strategy has not turned into a versatile tool in identifying the inter-molecular role in receptor binding sites through recorded activities by means of structurally selected common variables; that is to use QSAR information for internal mechanistic predictions among training molecules to see their inter-relation respecting the whole class of observed activities employed for a specific correlation. Such an approach will also be helpful for checking the chemical domain spanned by training molecules—a feature of the paramount importance also for further external tests.

The present communication wishes to start filling this gap by deepening the modeling of inter-molecular activity through extending the main concepts of recent developed SPECTRAL-SAR [32–40], developed the fully algebraic version of traditional statistically optimized QSAR picture, targeting the quantification of the competition between molecular inter-activity and inter-endpoints records.

QUASAR METHODOLOGY

Paradoxically, the main problem for QSAR resides not in performing the correlation itself but setting the variable selection for it; the mathematical counterpart for such problem is known as the "factor indeterminacy" [41-45] and affirms that the same degree of correlation may be reached with in principle an infinity of latent variable combinations. Fortunately, in chemical-physics there are a limited (although many enough) indicators to be considered with a clear-cut meaning in molecular structure that allows for rationale of reactivity and bindings [46, 47]. However, the main point is that given a set of N-molecules, one can chose to correlate their observed activities $A_{i=\overline{1,N}}$ with M-selected structural indicators in as many combinations as:

$$C = \sum_{k=1}^{M} C_M^k, \ C_M^k = \frac{M!}{k!(M-k)!},$$

(9.1a)

linked by different endpoint paths, as many as:

$$K = \prod_{k=1}^{M} C_M^k$$

(9.1b)

indexing the numbers of paths built from connected distinct models with orders (dimension of correlation) from $k = 1$ to $k = M$.

Basically, for each of the C-combinations a correlation (endpoint) QSAR equation is determined, say $Y_{l=\overline{1,C}} = \{y_l\}_{l=\overline{1,C}}^{i=\overline{1,N}}$, containing all computed activities for all considered N-molecules within the l-selected correlation. Now, the SPECTRAL-SAR version of QSAR analysis computes these activities in a complete non-statistical way, that is, by assuming the vectors for both observed (activities) and unobserved (latent variables) quantities while furnishing their correlation throughout a specific SPECTRAL-SAR determinant obtained from the transformation matrix between the orthogonal (desirable) and oblique (input) correlations. Yet, besides producing essentially the same results as the statistical least-square fit of residues the SPECTRAL-SAR method introduces new concepts as:

- *endpoint spectral norm*

$$\||Y_l\rangle\| = \sqrt{\langle Y_l|Y_l\rangle} = \sqrt{\sum_{i=1}^{N} y_{il}^2}, \ l = \overline{1,C} \tag{9.2}$$

allowing the possibility of the unique assignment of a number to a specific type of correlation, that is, performing a sort of resumed quantification of the models [34];

- *algebraic correlation factor*

$$R_{ALG,l} = \frac{\||Y_l\rangle\|}{\||A\rangle\|} = \sqrt{\frac{\sum_{i=1}^{N} y_{il}^2}{\sum_{i=1}^{N} A_i^2}}, \ l = \overline{1,C}, \tag{9.3}$$

viewed as the ratio of the spectral norm of the predicted activity to that of the measured one, giving the measure of the overall (or summed up) potency of the computed activities respecting the observed one rather than the local (individual) molecular distribution of activities around the mean statistical yields; thus, it is a specific measure of the molecular selection under study, always with a superior value to that yielded from statistical approach [37], however, preserving the same hierarchy in a shrink (less dispersive) manner being therefore better suited for intra-training set molecular analysis;

- *spectral path*, with the distance defined in the Euclidian sense as:

$$[l,l'] = \sqrt{\left(\||Y_l\rangle\| - \||Y_{l'}\rangle\|\right)^2 + (R_l - R_{l'})^2}, \ \forall(l,l') = \overline{1,C} \tag{9.4}$$

allows for defining complex information as path distances in norm-correlation space with norms computed from equation (9.2) while correlation free to be considered either from statistical (local) or algebraically (global)—equation (9.3) approaches; note that as far as computed activity Y_l corresponds to the measured activity A_l defined as logarithm of inverse of EC_{50}, see below, both modulus of Y_l vectors and R values have no units so assuring the consistency of the equation (9.4).

- *least spectral path principle*, formally shaped as:

$$\delta\left[l_1,...l_k...,l_M\right]=0;\ \ l_1,...,l_k,...,l_M:ENDPOINTS \tag{9.5}$$

provides a practical tool in deciding the dominant $\{\alpha,...\}$ hierarchies along the paths constructed by linking all possible k-models (i.e., models with k correlation factors) from (9.1a) combinations selected one time each on a formed path—generating the so called "M-endpoints containing ergodic path on K-paths assembly" of (9.1b). However, the implementation of the principle (9.5) is recursively performed through selecting the least distance computed upon systematically application of equation (9.4) on ergodic paths; if, by instance, two paths are equal there is selected that one containing the first two models with shorter norm difference in accordance with the natural least action; the procedure is repeated until all C-models where connected on shortest paths; there was already conjectured that only the first M-shortest paths (called as $\alpha_1,...,\alpha_M$) are enough to be considered for a comprehensive (and self-consistent) mechanistic analysis [34–40].

Nevertheless, for present purpose another two quantities are here introduced, namely:

• *inter-endpoint norm difference (IEND)*,

$$\Delta Y_{ll'} = \left\|\left|Y_{l'}\right\rangle\right\| - \left\|\left|Y_l\right\rangle\right\|,\ (l,l') \in \{\alpha_1,...,\alpha_M\} \tag{9.6}$$

that accounts for norm differences of the models lying on the M-shortest spectral paths linking M- from the C-models of equation (9.1a);

• *inter-endpoint molecular activity difference (IEMAD)*,

$$\Delta A_{i|j}^{l|l'} = A_j^{l'} - A_i^l = \ln\frac{1}{\left(EC_{50}\right)_j^{l'}} - \ln\frac{1}{\left(EC_{50}\right)_i^l} = \ln\frac{\left(EC_{50}\right)_i^l}{\left(EC_{50}\right)_j^{l'}} \tag{9.7}$$

is considered from activity difference between the fittest molecules (i, j), in the sense of minimum residues, for the models (l, l') belonging to the shortest paths $\alpha_1,...,\alpha_M$ for which the inter-endpoint norm difference is given by equation (9.6).

This way, we can interpret the two fittest molecules (i, j) as reciprocally activated by the models (l, l') through the spectral path whom they belong; put in analytical terms, the difference between quantities of equations (9.6) and (9.7) may assure the "jump" or *transition activity* that turns the effect of i molecule on that of j molecule across the least spectral (here revealed as metabolization) path connecting the models l and l':

$$\ln\frac{1}{q_{i|j}^{l|l'}} \equiv \Delta Y_{ll'} - \Delta A_{i|j}^{l|l'}. \tag{9.8}$$

Note that if we rearrange equation (9.8) in terms of EC_{50} of equation (9.7) one gets the wave-like form of molecular EC_{50} inter-molecular transformation:

$$\left(EC_{50}\right)_i^l = \left(EC_{50}\right)_j^{l'} q_{i|j}^{l|l'} \exp\left(i\Delta Y_{ll'}\right) \tag{9.9}$$

providing the analytic continuation in the complex plane for the *IEND* of equation (9.6) was assumed, that is, $\Delta Y_{i|l'} \rightarrow i\Delta Y_{i|l'}$ outside the factor $q_{i|j}^{l|l'}$. Remark that although the differences in equations (9.6) and (9.7) were consider mathematically, along the "arrow" *i-to-j*, the "quantum transformation" from equation (9.9) suggests that the bio-chemical-physical equivalence (metabolization) of the concentration effects evolves *from-j-to-i*, revealing a typical quantum behavior with the factor $q_{i|j}^{l|l'}$ playing the propagator role as the quantum kernels in path integral formulation of quantum mechanics [48].

Equation (9.9) stands as the present "quantum"-SAR/QUANTUM-SAR equation since:

- it involves *the wave-type* expression of molecular effect of concentration, however, for special selected molecules (the fittest out of the *C*-models) and for special selected paths (the least for the *M*-ergodic assembly), being *M* and *C* related by equation (9.1a);

- it provides the *specific transition* or specific transformation of the effect of a certain molecule into the effect of another special molecule out from the *N*-trained molecules, paralleling the phenomenology of consecrated quantum transitions;

- it has the amplitude of transformation driven by the so called *QUANTUM-SAR factor* of an exponential form

$$q_{i|j}^{l|l'} = \exp\left(\Delta A_{i|j}^{l|l'} - \Delta Y_{l|l'}\right) \tag{9.10}$$

defining the specific QUANTUM-SAR wave;

- it allows the *identity*

$$\left(EC_{50}\right)_i^l = \left(EC_{50}\right)_i^l \tag{9.11}$$

when the reverse effects is considered

$$\left(EC_{50}\right)_j^{l'} = \left(EC_{50}\right)_i^l \frac{1}{q_{i|j}^{l|l'}} \exp\left(-i\Delta Y_{l|l'}\right) \tag{9.12}$$

and substituted in the direct one (9.9), as absorption and emissions stand as reciprocal quantum effects;

- it has a "phase" with unity norm, in the same manner as ordinary quantum wave functions, allowing the inter-molecular *"real" QUANTUM-SAR transformation*

$$\left|\left(EC_{50}\right)_i^l\right| = q_{i|j}^{l|l'} \cdot \left|\left(EC_{50}\right)_j^{l'}\right| \tag{9.13}$$

exclusively regulated by the QUANTUM-SAR factor of equation (9.10), in the same fashion as quantum tunneling is characterized by the transmission coefficient;

- when *multiple transformations* take place across paths with multiple linked models, say (l, l', l''), the inter-molecular transformation $i{\rightarrow}j{\rightarrow}t$ is characterized by the overall QUANTUM-SAR factor (9.10) written as product of intermediary ones

$$q_{i|t}^{l|l''} = q_{i|j}^{l|l'} \cdot q_{j|t}^{l'|l''} \tag{9.14}$$

due to the two-equivalent ways the $(EC_{50})_i^l$ effect may be described directly from t or intermediated by j molecular effect transformations, respectively:

$$\begin{aligned}\left|(EC_{50})_i^l\right| &= q_{i|t}^{l|l''} \cdot \left|(EC_{50})_t^{l''}\right| \\ &= q_{i|j}^{l|l'} \cdot \left|(EC_{50})_j^{l'}\right| = q_{i|j}^{l|l'} \cdot \left(q_{j|t}^{l'|l''} \cdot \left|(EC_{50})_t^{l''}\right|\right)\end{aligned} \tag{9.15}$$

in the same way as the quantum propagators behave along quantum paths [48]; certainly, such contraction scheme may be generalized for least paths connecting the M-contained k-endpoints giving an overall QUANTUM-SAR ("metabolization power") factor as:

$$q_{i_1|i_M}^{l_1|l_M} = \prod_{w=2}^{M} q_{i_{w-1}|i_w}^{l_{w-1}|l_w} \tag{9.16}$$

- Equation (9.9) supports the *self-transformation* as well, with the driven qua-SAR factor given by:

$$q_{i|j=i}^{l|l'} = \exp\left(-\Delta Y_{l|l'}\right) \tag{9.17}$$

during its evolution along the least paths when the same molecule ($i=j$) is metabolized by activating certain structural features ($l \neq l'$) though specific indicators (variables) in correlation (bindings with receptor site); this case resembles the stationary quantum case according which even isolated (or with free motion), the molecular structures suffer dynamical wave-corpuscular or fluctuant transformation along their quantum paths.

With the present Qua-SAR methodology one can appropriately identify the molecular pairs that drive certain bio-/eco-activities against given receptor by means of selected descriptors in a "wave" or "quantum" mechanistic formal way. The ultimate goal will be the computation of QUANTUM-SAR factors along the least paths of actions that give the potential information of the conversion power of the fittest molecules in their specific bindings.

However, in order to practically understand the actual Qua-SAR approach all steps above will be in next specialized through an application for identifying the most involved polyphenolic molecules for their activity related to mammalian breast cancer.

APPLICATION TO FLAVONOIDS' ANTICANCER BIOACTIVITY

Although in general considered beneficial for their protective role in many age-related diseases - flavonoids (see Figure 9.1—with the general scheme in no.0) should be more carefully studied since their pharmacokinetics are not entirely elucidated [49–54].

For instance, recently, it was inferred that for certain flavonoids such as chrysin, nbiochanin A, and apigenin a very low micromolar concentration is capable of producing 50% (EC_{50}) of the maximum increase in mitoxantrone (MX) inhibitor substrate accumulation (interaction) with BCRP, helping in reversing the multidrug resistance (MDR) mechanism of overexpressing MCF-7 MX100 cancer cells [51–54].

Therefore, in order to assess the molecular role and structural- related mechanisms for potential lead compounds in the drug design for anti-cancer treatment, a series of representative classes of flavonoids have been employed, see Figure 9.1, with their recorded biological activities (A) among the computed transport (hydrophobicity-LogP), the electrostatic (polarizability POL), and steric (total energy at optimized 3D-configuration E_{TOT}) Hansch correlation variables [56], see Table 9.1, to successively provide the QSAR, SPECTRAL-SAR and finally to unfold the Qua-SAR analysis.

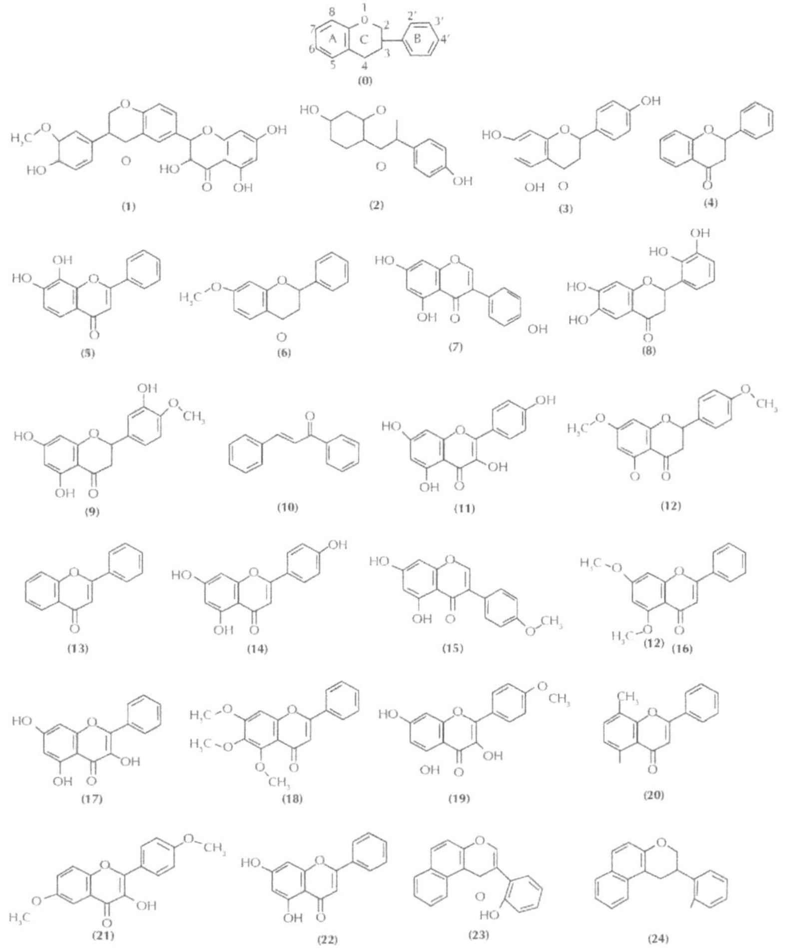

Figure 9.1. The studied flavonoids (with basic structure of as no.0 while the others are in the Table 9.1 characterized by associate QSAR data), covering the flavones, isoflavones, chalcones, flavonols, and flavanones, as they assist the increase of MX accumulation in BCRP-overexpressing MCF-7 MX100 breast cancer cells [51].

Table 9.1. The flavonoids of Figure 9.1 arranged by their ascending observed activities, defined as A = -log$_{10}$(EC$_{50}$[μM]) [51], along the associate computed structural parameters like the hydrophobicity (LogP), electronic cloud polarizability (POL) and the ground state configurationally optimized total energy (E$_{TOT}$) [55].

No.	Molecular Name	Activity	Structural parameters		
		A	LogP	POL(Å^3)	E$_{TOT}$(kcal/mol)
(1)	Silybin	3.74	2.03	45.68	−146625.1875
(9.2)	Daidzein	4.24	1.78	26.63	− 76984.7109
(9.3)	Naringenin	4.49	1.99	27.46	− 85032.9218
(9.4)	Flavanone	4.6	2.84	25.55	− 62849.3125
(9.5)	7,8-Dihydroxyflavone	4.7	1.75	26.63	− 76982.1328
(9.6)	7–Methoxyflavanone	4.79	2.59	28.02	− 73823.8046
(9.7)	Genistein	4.83	1.50	27.27	− 84380.7578
(9.8)	6,2',3'-7-Hydroxyflavanone	4.85	1.70	28.10	− 92422.6640
(9.9)	Hesperetin	4.91	1.73	29.93	− 96003.9921
(9.10)	Chalcone	4.93	3.68	25.49	− 55450.1093
(9.11)	Kaempferol	5.22	0.56	27.90	− 91770.5859
(9.12)	4'-5,7-Trimethoxyflavanone	5.25	2.08	32.96	− 95768.9062
(9.13)	Flavone	5.4	2.32	25.36	− 62196.3437
(9.14)	Apigenin	5.78	1.46	27.27	− 84379.8593
(9.15)	Biochanin A	5.79	1.53	29.10	− 87961.2812
(9.16)	5,7-Dimethoxyflavone	5.85	1.81	30.30	− 84139.4687
(9.17)	Galangin	5.92	0.85	27.27	− 84376.8359
(18)	5,6,7-Trimethoxyflavone	5.96	1.56	32.77	− 94976.1875
(19)	Kaempferide	5.99	0.60	29.74	− 95351.3984
(20)	8-Methylflavone	6.21	2.79	27.19	− 65789.9218
(21)	6,4'-Dimethoxy-3-hydroxy-flavone	6.35	0.41	31.13	− 92162.7187
(22)	Chrysin	6.41	1.75	26.63	− 76986.1171
(23)	2'-Hydroxy-α-naphtoflavone	7.03	3.07	33.26	− 82027.8359
(24)	7,8-Benzoflavone	7.14	3.35	32.63	− 74634.5234

Note that in Table 9.1 the molecules were displayed in ascendant order of their recorded activities, from no. 1–24, for having present which is superior to which each time they are reciprocally quotation. Such an arrangement allows the construction of an activity differences chart, see Table 9.2, with great utility in establishing *the inter-endpoint molecular activity differences* of equation (9.7) entering *QUANTUM-SAR factor* of equation (9.10). Next, for computing the other influential activity difference in Qua-SAR, namely the *inter-endpoint norm difference* of equation (9.6), the C = 10 possible endpoint models with data of Table 9.1 are in Table 9.3 presented. However, worth remarking that the traditional hydrophobicity factor LogP seems to have quite little or even no-influence from traditional statistical correlation (model *Ia*).

Table 9.2. The anti-symmetric matrix of the inter-molecular activity differences for the working flavonoids of Table 9.1.

1	2	3	4	5	6	7	8	9	10	11	12	13	14	15	16	17	18	19	20	21	22	23	24	
0	0.5	0.75	0.86	0.96	1.05	1.09	1.11	1.17	1.19	1.48	1.51	1.66	2.04	2.05	2.11	2.18	2.22	2.25	2.47	2.61	2.67	3.29	3.4	*1*
	0	0.25	0.36	0.46	0.55	0.59	0.61	0.67	0.69	0.98	1.01	1.16	1.54	1.55	1.61	1.68	1.72	1.75	1.97	2.11	2.17	2.79	2.9	*2*
		0	0.11	0.21	0.3	0.34	0.36	0.42	0.44	0.73	0.76	0.91	1.29	1.3	1.36	1.43	1.47	1.5	1.72	1.86	1.92	2.54	2.65	*3*
			0	0.1	0.19	0.23	0.25	0.31	0.33	0.62	0.65	0.8	1.18	1.19	1.25	1.32	1.36	1.39	1.61	1.75	1.81	2.43	2.54	*4*
				0	0.09	0.13	0.15	0.21	0.23	0.52	0.55	0.7	1.08	1.09	1.15	1.22	1.26	1.29	1.51	1.65	1.71	2.33	2.44	*5*
					0	0.04	0.06	0.12	0.14	0.43	0.46	0.61	0.99	1	1.06	1.13	1.17	1.2	1.42	1.56	1.62	2.24	2.35	*6*
						0	0.02	0.08	0.1	0.39	0.42	0.57	0.95	0.96	1.02	1.09	1.13	1.16	1.38	1.52	1.58	2.2	2.31	*7*
							0	0.06	0.08	0.37	0.4	0.55	0.93	0.94	1	1.07	1.11	1.14	1.36	1.5	1.56	2.18	2.29	*8*
								0	0.02	0.31	0.34	0.49	0.87	0.88	0.94	1.01	1.05	1.08	1.3	1.44	1.5	2.12	2.23	*9*
									0	0.29	0.32	0.47	0.85	0.86	0.92	0.99	1.03	1.06	1.28	1.42	1.48	2.1	2.21	*10*
										0	0.03	0.18	0.56	0.57	0.63	0.7	0.74	0.77	0.99	1.13	1.19	1.81	1.92	*11*
											0	0.15	0.53	0.54	0.6	0.67	0.71	0.74	0.96	1.1	1.16	1.78	1.89	*12*
												0	0.38	0.39	0.45	0.52	0.56	0.59	0.81	0.95	1.01	1.63	1.74	*13*
													0	0.01	0.07	0.14	0.18	0.21	0.43	0.57	0.63	1.25	1.36	*14*
														0	0.06	0.13	0.17	0.2	0.42	0.56	0.62	1.24	1.35	*15*
															0	0.07	0.11	0.14	0.36	0.5	0.56	1.18	1.29	*16*
																0	0.04	0.07	0.29	0.43	0.49	1.11	1.22	*17*
																	0	0.03	0.25	0.39	0.45	1.07	1.18	*18*
																		0	0.22	0.36	0.42	1.04	1.15	*19*
																			0	0.14	0.2	0.82	0.93	*20*
																				0	0.06	0.68	0.79	*21*
																					0	0.62	0.73	*22*
																						0	0.11	*23*
																							0	*24*

The first conclusion is that flavonoids have practically no exclusive or primarily role in drug transporting to BCRP site; still, the electrostatic influence through POL is practically missing as well (model *Ib*), while the stericity through ETOT unfolds some statistically sensitive role in ligand (MX)-receptor (BCRP) binding (model *Ic*). The last assertion may also be sustained by going to the two-correlated parameters end-point models, when one can see the confirmation of the stericity role through ETOT correlation variable: while combination LogP∧POL does not improve the statistical correlation of model *IIa* significantly over single-parameter LogP∨POL correlations, the total energy presence provides better and better correlation behavior as it is combined with LogP (the model *IIb*) and with POL (the model *IIc*), respectively. Instead, when all the Hansch structural variables are taken into account the model *III* is generated with appreciable statistical correlation respecting the other computed combinations.

Overall, it cannot be inferred that LogP and POL does have no influence on correlation only because when alone they do not correlate at all with flavonoids' bioactivity, because their cumulative presence in model *III* highly improves the single E_{TOT} correlation of model *Ic* as well as mixed correlations of bi-variable models *IIb* and *IIc*. Therefore, the mechanistic "alchemy" of structural features on molecular activity seems complex enough when all hydrophobicity, electrostatic and stericity influences combine as they are reciprocally activating one each other with a superior resultant in modeling ligand-receptor binding.

Yet, the algebraic correlation factors in Table 9.3 deserve special discussion: it is clear that as they are not measuring the dispersive character of the local computed (molecular) points against the average recorded activity as statistical metrics do, their values are all close to unity and close to each other as well; however, they are modeling another reality of computation, being closer to path integral approach than to differential analysis, through indexing the global behavior or the total length of the computed vector to the recorded one. Still, while between the algebraic and statistical correlations only an indirect connection exists [37], the one-to-one hierarchical ordering of models is always recorded thus supporting the usefulness of using algebraically scale when the shrink of correlation factors is more favorable. For instance, in the present case, as above revealed, according to the statistical analysis, there seems that LogP (*Ia*) and POL (*Ib*) have no influence on correlation, while when combined with E_{TOT} in model *III* they considerably enrich the single E_{TOT} correlation power of model *Ic*. Such behavior shows that orthogonal, that is, independent, descriptors may provide better results when are combined than when considered apart due to the increase of the (inter) correlation space.

Having performed the QSAR analysis, the specific SPECTRAL-SAR stage can be unfolded by means of the ($K = 9$, $M = 3$) ergodic paths with the spectral Euclidian lengths given by equation (9.4) in both statistical and algebraic frameworks, as shown in Table 9.4. Next, the least $M = 3$ paths with the dominant M-factors influence are selected by applying the above exposed recursive rule of *least path principle* resumed by equation (9.5). Remarkably, there follows that the resulting alpha (a), beta (b), and gamma (g) most influential paths are identically shaped no matter whether statistical or algebraically schemes are undertaken. This result, although not necessarily viewed as a general rule, shows that in this specific case the algebraically analysis leaves with systematically the same mechanistically results as those obtained with statistical tools. However, once more, we stress on that algebraically measure may give more realistic inside in the Q(Spectral)-SAR phenomenology since its inner vectorial and norm-based algorithm accounting for each individual molecular contribution to the whole activity "basin" rather than respecting the average activity.

Table 9.3. QSAR equations through SPECTRAL-SAR multi-linear procedure [32–34] for all possible correlation models considered from data of Table 9.1; here $|X_0\rangle$ is the unitary vector $|11...1_{24}\rangle$, while the structural variables are set as $|X_1\rangle = LogP$, $|X_2\rangle = POL$, and $|X_3\rangle = E_{TOT}$; the predicted activities' norms where calculated with equation (9.2), while the algebraic correlation factor of equation (9.3) uses the measured activity of $\|A\rangle\| = 26.9357$ computed upon equation (9.2) with data of Table 9.1; $R_{Statistic}$ is the traditional Pearson correlation factor [1–8].

Model	Variables	(Q/S-)SAR Equation	$\|Y\rangle^{PREDICTED}\|$	$R_{Algebraic}$	$R_{Statistic}$
Ia	$|X_0\rangle, |X_1\rangle$	$|Y\rangle^{Ia} = 5.39837|X_0\rangle+0.0179106|X_1\rangle$	26.6138	0.988049	0.0175601
Ib	$|X_0\rangle, |X_2\rangle$	$|Y\rangle^{Ib} = 5.67735|X_0\rangle-0.00834411|X_2\rangle$	26.61425	0.988065	0.0409922
Ic	$|X_0\rangle, |X_3\rangle$	$|Y\rangle^{Ia}=6.48303|X_0\rangle+0.0000124625|X_3\rangle$	26.6344	0.988812	0.252513
IIa	$|X_0\rangle,$	$|Y\rangle^{IIa} = 5.64318|X_0\rangle$	26.614349	0.988069	0.0445618
	$|X_1\rangle, |X_2\rangle$	$+0.0178242|X_1\rangle-0.00833676|X_2\rangle$			
IIb	$|X_0\rangle,$	$|Y\rangle^{IIb} = 6.93331|X_0\rangle$	26.638	0.988947	0.273909
	$|X_1\rangle,|X_3\rangle$	$-0.120924|X_1\rangle+0.0000150708|X_3\rangle$			
IIc	$|X_0\rangle,$	$|Y\rangle^{IIc} = 4.99884|X_0\rangle$	26.6681	0.990063	0.409837
	$|X_2\rangle,|X_3\rangle$	$+0.122989|X_2\rangle+0.0000376701|X_3\rangle$			
III	$|X_0\rangle,$	$|Y\rangle^{III} = 5.59424|X_0\rangle$	26.7758	0.994064	0.708509
	$|X_1\rangle,|X_2\rangle,$	$-1.05993|X_1\rangle+0.400704|X_2\rangle$			
	$|X_3\rangle$	$+0.000117452|X_3\rangle$			

Table 9.4. Synopsis of paths connecting the endpoints of Table 9.3 in the norm-correlation spectral-space.

Path	Value	
	Algebraic	**Statistic**
Ia-IIa-III	0.162142	0.710311
Ia-IIb-III	0.162142	0.713422
Ia-IIc-III	γ0.162142	γ0.713533
Ib-IIa-III	β0.161697	β0.686875
Ib-IIb-III	0.161697	0.690271
Ib-IIc-III	0.161697	0.690059
Ic-IIa-III	0.181617	0.892215
Ic-IIb-III	α0.141579	α0.477638
Ic-IIc-III	0.141579	0.478416

Going now to the individual molecular level analysis, Table 9.5 lists the residual activities between computed and observed activities for each of considered models, distributed along the already identified least paths. At this instance, the most fitted molecule is outlined out of each endpoint; most impressive, the actual research selected the same molecule as the best fitted one along the both a and b paths, namely molecule no. 12 (4'-5,7-trimethoxyflavanone) and molecule no. 13 (flavone), respectively. Moreover, these molecules are not among the most potent one respecting the observed activity of Table 9.1, being situated at the middle to second-half panel of the 24 molecules considered.

Table 9.5. Residual activities $A_i–Y_i^{Model}$ of the compounds of Table 9.1 for the SPECTRAL-SAR models of Table 9.3 ordered according with the alpha, beta, and gamma paths of Table 9.4; that residue which is closes to zero in each considered endpoint is marked by a line border.

No.	Models						
	α		β		γ		
	Ic	IIb	Ib	IIa	Ia	IIc	III
1	−0.915706	−0.738065	−1.5562	−1.55854	−1.69473	−1.35359	−0.785284
2	−1.2836	−1.31784	−1.21515	−1.2129	−1.19025	−1.13401	−1.09626
3	−0.933302	−0.921149	−0.958225	−0.959719	−0.944015	−0.682916	−0.0110057
4	−1.09977	−1.04269	−0.864162	−0.880793	−0.849239	−1.17367	−0.840236
5	−0.823636	−0.861503	−0.755151	−0.752361	−0.729716	−0.674109	−0.668387
6	−0.772996	−0.717525	−0.653552	−0.665745	−0.654761	−0.874038	−0.615978
7	−0.60143	−0.650231	−0.619811	−0.612569	−0.595239	−0.344115	−0.190835
8	−0.481207	−0.484848	−0.592885	−0.589214	−0.578821	−0.123257	0.653106
9	−0.376575	−0.367246	−0.517615	−0.514493	−0.519358	−0.153417	0.432251
10	−0.86198	−0.722625	−0.534663	−0.566265	−0.534284	−1.11502	−0.464907
11	−0.119334	−0.262529	−0.224554	−0.200562	−0.188403	0.246777	−0.181659
12	−0.0395047	0.0115343	−0.152333	−0.155471	−0.185627	−0.194929	−0.098518
13	−0.307904	−0.31541	−0.0657478	−0.0731081	−0.0399254	−0.374895	−0.591958
14	0.348559	0.294919	0.330189	0.338144	0.355478	0.605851	0.716663
15	0.403192	0.367358	0.355459	0.362152	0.364224	0.525694	0.488215
16	0.415563	0.403619	0.425472	0.427166	0.419209	0.294139	−0.0847295
17	0.488521	0.361109	0.470189	0.489017	0.506403	0.745737	0.209751
18	0.660616	0.646707	0.556082	0.562214	0.533687	0.508577	0.0433466
19	0.695292	0.566274	0.560799	0.584065	0.580881	0.925367	0.314017
20	0.546881	0.605582	0.759522	0.743771	0.761657	0.345406	0.404994
21	1.01555	0.855242	0.932398	0.959039	0.944284	0.994295	−0.458865
22	0.886414	0.848557	0.954849	0.957639	0.980284	1.03604	1.04208
23	1.56925	1.70416	1.63017	1.60938	1.57664	1.03055	0.996681
24	1.58711	1.7366	1.73491	1.70914	1.68163	0.939523	0.787543

Such result tells us that the maximum recorded activity is not necessarily that one induced by *specific* chosen structural variables (here as LogP, POL, and E_{TOT}). This is the case of the most fitted molecule on the most correlated endpoint (III) appeared to be no.3 (naringenin), with low activity on the observed range compared with the no. 25 (7,8-benzoflavone) in Table 9.1. Consequently, one can say that the first half of the observed activities in Table 9.1 may be attributed to certain physico-chemical indicators with clear mechanistically roles, while the rest of observed activities may be due to other unidentified specific structural descriptors or even to non-specific ones (rooting in the sub-quantum nature of the particular observer-observed system). Nevertheless, this lower activity prescribed by the computational results is in accordance with the so called "homeopathic principle" prescribing cure by moderate-to-low active drugs while better monitoring their effects through controlled physico-chemical descriptors.

For the sake of comparison, the actual Spectral(Qua)SAR results are to be compared with the consecrated PCA [57]. This way, Figure 9.2 illustrates the graphical 3D correlations among the descriptors LogP, POL, and E_{TOT} used in this study; it offers a visual way for assessing the almost no-correlation of LogP with other concerned variables, POL and E_{TOT}, respectively. This lead with conclusion that LogP is almost orthogonal (independent) on (respecting) the other two Hansch variables. Instead, when further performing the factor analysis, the Table 9.6 is obtained while clearly revealing the scarce correlation carried by considering LogP variable alone. This is in close agreement with the SPECTRAL-SAR results, see above. In any case, the hydrophobicity description and its descriptor cannot be rejected only by factor analysis since it drives (firstly or latter) the inter-membrane interaction that is essential for drug-cell binding. Spectral- and Qua-SAR highly proved the important role hydrophobicity plays in combination with electrostatic (POL) and steric (E_{TOT}) interactions. Moreover, while PCA shows the POL factor influence equals that of E_{TOT}, whereas their role in correlation is sensible different in SPECTRAL-SAR analysis (compare model *Ib*-last column of Table 9.3 with POL-last column of Table 9.6). However, again, this discrepancy is in the favor of SPECTRAL-SAR since the PCA results are due to the sensitive degree of POL-E_{TOT} correlation (see Figure 9.2), from where the PCA yield that POL and E_{TOT} display similar correlation power, while SPECTRAL-SAR includes also the orthogonalization of POL and E_{TOT} variables prior correlation takes effect and better discriminates among their influence in bonding.

Nevertheless, going ahead with the SPECTRAL-SAR results the Qua-SAR factors may be immediately recover by employing the molecular activity differences from Table 9.2 for the best fitted molecules of Table 9.5 along the models of the most influential paths in molecular mechanism towards MX-BCRP binding. The resulted *IEND* and *IEMAD* of Table 9.7 are combined to produce the QUANTUM-SAR factors of equation (9.10) type for each two-molecules-two-models on specific paths, while the "metabolization power" *per* path is finally obtained by their couplings, according with multiplicative quantum rule of equation (9.16). Worth noting that the overall QUANTUM-SAR factors of paths are in total agreement with the previous SPECTRAL-SAR selected path hierarchy, that is, the a path is associated with the highest q-SAR factor, being followed by that of b path and by that of g one in last column of Table 9.7.

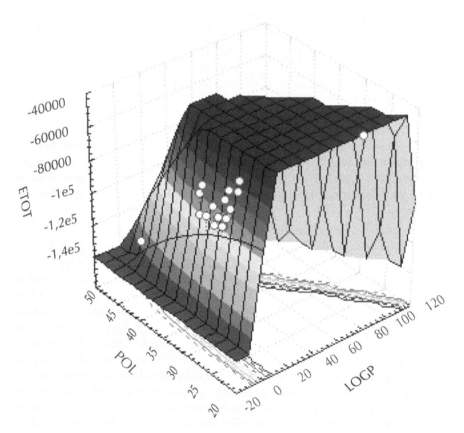

Figure 9.2. Quadratic 3D representation of LogP *vs.* POL *vs.* E_{TOT} variables' fit employing the data of Table 9.1 [58].

Table 9.6. Principal Component Analysis (PCA) for the data of Table 9.1 within unrotated (unnormalized) factor score coefficients [58].

	PC1	*PC2*	*PC3*	*Multiple PC1-PC3 factors' R²*
Eigenvalue:	**1.958158**	**0.892127**	**0.149715**	
% total variance:	**65.27195**	**29.73757**	**4.99049**	
Variable	*Factors' coefficients*			
LogP	0.232179	−0.997780	0.20467	0.083712
POL	−0.472177	−0.302902	−1.79349	0.716820
E_{TOT}	0.483556	0.183309	−1.84956	0.728872

Table 9.7. Determination of the QUANTUM-SAR, see equation (9.10) with equations (9.6) and (9.7), associate with certain couple of molecules involved in activating specific structural quantum indices (or their combinations) driving spectral paths of Table 9.4, by employing minimum residue recipe throughout Table 9.5 for each considered endpoint, as well as the associate recorded bioactivity differences of Table 9.2, respectively.

Path	$\Delta Y_{t|t'}^{PATH}$ (IEND)#	$\Delta A_{i|j}^{l|l'}$ (IEMAD)♦	$q_{i|j}^{l|l';PATH}$ *	q^{PATH}
α	$\Delta Y_{Ic|IIb}^{\alpha}=0.00364573$	$A_{12|12}^{Ic|IIb}=0$	$q_{12|12}^{Ic|IIb;\alpha}=0.991641$	$q^{\alpha}=0.125464$
	$\Delta Y_{IIb|III}^{\alpha}=0.137836$	$A_{12|3}^{IIb|III}=-0.76$	$q_{12|3}^{IIb|III;\alpha}=0.126521$	
β	$\Delta Y_{Ib|IIa}^{\beta}=0.0000989324$	$A_{13|13}^{Ib|IIa}=0$	$q_{13|13}^{Ib|IIa;\beta}=0.999772$	$q^{\beta}=0.0848036$
	$\Delta Y_{IIa|III}^{\beta}=0.161487$	$A_{13|3}^{IIa|III}=-0.91$	$q_{13|3}^{IIa|III;\beta}=0.0848229$	
γ	$\Delta Y_{Ia|IIc}^{\gamma}=0.0542592$	$A_{13|8}^{Ia|IIc}=-0.55$	$q_{13|8}^{Ia|IIc;\gamma}=0.248737$	$q^{\gamma}=0.0847168$
	$\Delta Y_{Ia|IIc}^{\gamma}=0.107771$	$A_{8|3}^{IIc|III}=-0.36$	$q_{8|3}^{IIc|III;\gamma}=0.340588$	

#Inter-Endpoint Norm Difference, equation (9.6); ♦Inter-Endpoint Molecular Activity Difference, equation (9.7);

*Note that here the basic relation of equation (9.10) was considered in decimal base since originally, the associated activities in Table 9.1 were as such defined.

This result may be quite important if such a behavior may be proven to hold in general since it would allow the effective quantification of paths according with their metabolization power. However, such endeavor exceeds the present communication purpose and will remain as a future challenge in Qua-SAR studies.

Finally, all QSAR, SPECTRAL-SAR and Qua-SAR computational results may be collected and resumed by associate "spectral" scheme for evolution of the fittest molecular structures along the endpoint models for the (M=3) selected mechanistic paths of actions, see Figure 9.3. Note that algebraic correlation environment was chose as the "vertical" indicator for the degree with which a certain model reaches the observed activity in the vectorial norm sense (equally, the norm themselves could be used for ordinate axis [34]).

Going now to comment upon the "metabolization power" as indicated by the QUANTUM-SAR factors on Figure 9.3, one can firstly observe that for the a path the "first movement" from the Ic (E_{TOT}) to IIb ($LogP \wedge E_{TOT}$) corresponds to quantum free motion so that the null IEMAD for molecule no. 12 (4'-5,7-trimethoxyflavanone) is carried; here, the quantum metabolization factor $q_{12|3}^{IIb|III;\alpha}$ is consumed only for strongly activating the membrane transporter feature (LogP) of the same molecule. Instead, on the last passage of the a path the factor $q_{12|12}^{Ic|IIb;\alpha}$ is responsible for converting the electrostatic (POL) influence of the flavonoids no. 12 towards no.3 (naringenin) activity as well as for reverse-O-methylation (methoxylation) of oxygens in positions 5 and 7 (on ring A) and 4' (on ring B) respecting the molecular pattern no.0 in Figure 9.1, respectively. Such result is in fully accordance with the reverse quantum influence that is at the foreground of QUANTUM-SAR factor conversion prescribed by equation (9.9), that is, quantifying the power of back transformation of molecular EC50s respecting the "arrows" of IEND and IEMAND in equation (9.10). However, the fact that such transformation is the first one acting at molecular level is sustained also by optimized 3D configurations of involved molecules no. 12 and 13, being both with rings A and B spatially bent in Figure 9.3 respecting the ring C of the planar pattern no.0 of Figure 9.1.

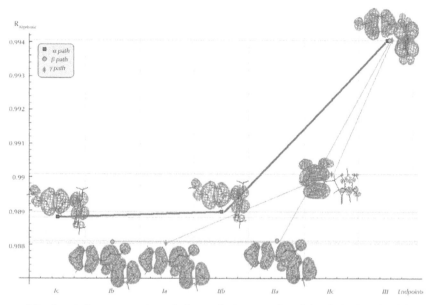

Figure 9.3. Spectral representation of the endpoints employed in designing the bioactivity mechanism for the molecules of Table 9.1, according with the algebraic correlation factors of equation (9.3) in Table 9.3, across the shortest (three) paths identified from Table 9.4, while marking the fittest molecules' orbital 3D-distribution for each considered model, that is, molecule no. 12 (4'-5,7-trimethoxyflavanone) for the models *Ic* and *IIb*, molecule no. 13 (Flavone) for models *Ib*, *Ia*, and *IIa*, molecule no. 8 (6,2',3'-7-tydroxyflavanone) for model *IIc*, and molecule no. 3 (naringenin) for model *III*, respectively.

A somewhat different situation is met for b path in Figure 9.3; in its first part a higher q-SAR factor ($q_{13|3}^{Ib|IIa;\beta} > q_{12|12}^{Ic|IIb;\alpha}$) is needed for activating the transporter hydrophobicity feature in model *IIa* (LogP∧POL) starting from model *Ib* (POL), while in its second part the molecule no. 13 (flavone) is shown to be metabolized in molecule no. 3 (naringenin) by a direct hydroxylation in positions 5, 7 (on ring A), and 4' (on ring B), the same as before, respecting the molecular pattern no.0 in Figure 9.1, by a smaller q-SAR factor, $q_{13|3}^{IIa|III;\beta} << q_{12|3}^{IIb|III;\alpha}$, compared with that involved in the previous alpha path. Despite these, the overall quantum factor of beta path is lower than that of alpha, meaning a decrease capacity of metabolization since direct addition is involved, contrarily to the ordinary "inverse" QUANTUM-SAR transformation of equation (9.9), while stericity (here founded as the most influential QSAR variable) is triggered by more steric energy difference consumed between the planar optimized configuration of molecule no. 13 on that spatially bended of molecule no. 3 in Figure 9.3.

Even more metabolization "operations" take place along the g path of Figure 9.3: there is started on the same planar configuration of molecule no.13 (flavone); then, the q-SAR factor $q_{13|8}^{Ia|IIc;\gamma}$ turn it into molecule no. 8 (6,2',3'-7-hydroxyflavanone) by hydroxylation on the indicated positions (6 and 7 for ring A and 2' and 3' for ring B respecting the pattern molecule no.0 of Figure 9.1) while activating electrostatic and steric factors in model IIc ($POL \wedge E_{TOT}$) from independent hydrophobicity factor of model Ia (LogP)—a complex movement that explain why this molecular path comes at the final, with less probability and potency; nevertheless, this path has on its last passage no less complex transformation, that is, turning the molecule no. 8 into no. 3 one by combined reverse hydroxylation in positions 2' and 3' with direct hydroxylation of position 4' on B ring and with movement from ortho (9.6)–para (9.5) of hydroxyl group on ring A respecting pattern molecule no. 0 of Figure 9.1, respectively; the transformation efficiency $q_{8|3}^{IIc|III;\gamma}$ is a bit higher than on the first part of the path since it require less steric energy consumption to bent the ring C respecting the A-B ones while accounting for electronic delocalization density (orbitals) over them until the configuration of molecule no. 3 is reached in Figure 9.3.

Overall, it is clear that the Qua-SAR scheme offers a quantification recipe along the most effective spectral paths combined with most fitted molecules for a trial basin of analogues compounds and structural variables. In the present case there was revealed that the energetic steric factor E_{TOT} seems to mainly drive the mechanistic molecular transformation in MX-BCRP binding phenomenology, while the molecule no. 3 (naringenin) appears as the best fitted molecules belonging to the most relevant endpoint, in clear disjunction with the roughly molecular selection upon initial input observed activity data. That is, naringenin (no. 3) is shown to be the best adapted molecule for the actual Log $P \wedge POL \wedge E_{TOT}$ structural (independent) factors being metabolized from molecules as 6,2',3'-7-hydroxyflavanone (no. 8), 4'-5,7-trimethoxyflavanone (no. 12), and flavone (no.13) by specific molecular mechanistically paths. However, there appears that these molecules are not linked even through the paths with the most active compounds of Table 9.1; statistically, this can be explained by the so called "regression towards the mean" effects, in the sense that the best correlations translated to the compounds found in the middle of the mentioned sorted Table 9.1; from the structural point of view such behavior may attributed to the specific parameters used for correlations that best describe molecules with specific groups, most favorable for the descriptor's nature.

On the other hand, the present study affirms the position 7 of ring A and position 4' of ring B respecting the pattern molecule no. 0 of Figure 9.1 as the most suitable ones for producing an increase in BCRP inhibition activity, given that these positions belong to the a and b paths and being common to the rest of spectral paths as well. Instead, the position that does not appear at all in any of the a, b, or g paths, namely position 8 on ring B may present adverse drug interactions.

Further Qua-SAR studies are necessary and will be developed for exploring other bio- and eco-active compounds for their interactions with organs and organisms; they may hopefully lead to a coherent analytical picture of chemical–biological bonding focused on selecting the most adapted molecules and of the most privileged molecular positions for delivering controlled structural based chemical reactivity and biological activity.

CONCLUSION

The modern *in silico* (computational) chemical analysis respecting the bio-activity and availability of analogues substances, potentially beneficial or detrimental for specific interaction in organs and organisms, faces with a paradoxical dichotomy: if searching for the best correlation useful for *prediction* of specific molecular bio- or eco-activity QSAR models involving un-interpretable many latent variables may be produced, while always remaining the question of correlation factor indeterminacy (i.e., the assumed descriptors can be at any time replaced with other producing at least the same correlation performances); instead, when restricting the analysis to search for molecular design and mechanisms throughout performing SARs by means of special structural indicators for a given class of relevant molecules, arises the price of limiting the use of generated models for further prediction. The present communication is mainly devoted in developing the second (Q)SAR facet by extending the recent introduced notion of spectral-path-linking-endpoints and the associate least action principle to spectral path quantification, in terms of the best fitted molecules, along the contained computed models, by means of the introduced q(uantum)-SAR factor within the generally called QUANTUM-SAR (QuaSAR) methodology.

As an application, for representative flavonoids' inhibiting activities on breast cancer resistant protein there was clearly shown that the newly introduced q-SAR factor offers relevant analytical characterization of previously conceptually introduced spectral path hierarchy; moreover, the present QuaSAR may allow interpretation interconversion of concerned molecules' towards receptor binding since belonging to the same class of analogs, while they certainly undertaking such transformation during their interaction with macromolecules, proteins, and enzymes present on cellular walls or with *in vivo* environment.

Basically, the QuaSAR stands as the first step in assessing the quantum mechanically equivalent of wave function to the sample of molecules interacting with a specific organism site; it will eventually lead with the hyper-wave function with the help of which the associate hyper-density probability of binding (metabolization) is to be computed; the last information may provide the density probability map of the ligand-receptor interaction abstracted from the structural Spectral-Qua-SAR correlations; with this tool the molecular design of new chemical structures may be appropriately undertaken.

However, the actual QuaSAR scheme and quantum factor carry the main features of quantum dynamical systems and may stimulate future computational and conceptual developments in molecular design for structurally controlled activity. Further, generalization of the present QuaSAR method to modeling all potential inter-conversions of employed molecules involved in correlation as well as for establishing their quantum metabolization complete map (through, for instance, hydrophobic, electrostatic, and steric barrier tunneling) is actually in progress and will be reported in subsequent communications.

KEYWORDS

- **Ergodic**
- **Flavonoids**
- **Naringenin**
- **Stericity**

Chapter 10

About the SPECTRAL-SAR Overcome to Statistical Approach of QSAR: An Ecotoxicity Case of Aliphatic Amines

Mihai V. Putz, Ana-Maria Putz, Marius Lazea, and Adrian Chiriac

---

INTRODUCTION

In the context of quantitative structure-activity relationship (QSAR) studies the tradi-tional statistical indices, that is, simple correlation factor, standard error of estimates, as well as t-Student and Fisher tests are challenged against the algebraic spectral de-scription of the multi-linear models. While the new correlation approach is built on the algebraic (Euclidian) norms and on the associate correlation factor, the validation and predictive features are assured throughout so called "ergodic" *least action prin-ciple* across the investigated models. The reliability of the present proposed spectral respecting statistical QSAR analysis is evaluated on modeling the aliphatic amines' ecotoxicity by means of hydrophobicity, polarizability, and total energy structural pa-rameters resulting in better predicting of molecular hierarchy of observed bioactivity being susceptible of as such general behavior.In QSAR studies, the "goodness of the fit" and "prediction ability" issues have been distinguished as the most important mat-ters regarding the validation and usefulness of the provided models [1].

However, apart of the foreground conditions regarding the correlation data and structural parameter selection (i.e., the normal distribution of data [2], the symmetry/asymmetry problem of biological activities associate with active/inactive chemical compounds tested [3], the independence of the predictor variable avoiding multico-linearity [4], or the degrees of freedom for predicting variable such that the chance correlation to be reduced [5]) the QSAR validation philosophy encompasses two rela-tively disjoint trends: one advanced in performing a significant number of statistical tests such that to prevent the type I or errors (i.e., excluding a right hypothesis by con-sidering it false) [6], while the second recommends massive calculations under boot-strapping [7] and cross-validation analysis [8] in order to safely replace the standard statistical assumptions in certifying a relationship or model for minimizing the type II of errors (i.e., accepting a false hypothesis as being true) [9].

Generally, there is a difficult task to chose among above two approaches; nev-ertheless, by considering a QSAR validation more closely related with the physico-chemical structural parameters a sort of descriptive statistics beyond the "blind" tests upon a priori or selected model validation would leave with more oriented mechanisti-cally interpretation of "facts" causing observed activity. Yet, such a phenomenological QSAR analysis should imply the abolishment of the statistical indices cult for r, r^2, PRESS, q^2, and so forth [10].

In developing this line of multi-regression description, the recent studies introducing the so called SPECTRAL-SAR framework implying algebraic concepts as generalized scalar product, Gram–Schmidt orthogonalization procedure [11], Euclidian norms and paths have furnished the required tools for challenging the statistical validation of QSAR within alternative algebraic picture [12]. Such endeavor is in this study unfolded in its conceptual side firstly, being then illustrated throughout a computational application on structurally explained ecotoxicity of primary, secondary, and tertiary aliphatic amines [13].

ALGEBRAIC SPECTRAL-SAR APPROACH

Given a set of (biological) activities of N-compounds and a set of M-structural properties for each of them, one can accommodate both the activity data and the independent variables like N-dimensional vectors $|Y\rangle = |y_1 \quad y_2 \quad ... \quad y_N\rangle$ and $|X_l\rangle = |x_{1l} \quad x_{2l} \quad ... \quad x_N\rangle_{l=\overline{1,M}}$, respectively. In these conditions the vectorial QSAR regression will look like:

$$|Y\rangle = b_0|X_0\rangle + b_1|X_1\rangle + ... + b_k|X_k\rangle + ... + b_M|X_M\rangle \qquad (10.1)$$

with the correlation parameters $b_j, j = 0, ..., M$ while the unity vector $|X_0\rangle = |1 \quad 1 \quad ... \quad 1_N\rangle$ accounts for the free term coefficient b_0. This is the framework in which the so called algebraic SPECTRAL-SAR approach and analysis were provided. Worth noting that the vectorial transcription of the QSAR problem combined with Schmidt successive orthogonalization procedure of independent variables produces the orthogonal expansion of the given activity over the $(M+1)$-orthogonal generated vectors justifying this way the "spectral" appellative of the method, in the same manner as a given atomic or molecular property may be expressed in terms of the wave function discrete spectrum of the bounded systems. The transformation $(M+2)$-determinant between the initial N-dimensional space of data and the orthogonal-generated one yields the so called SPECTRAL-SAR equation that happens to can be algebraically rearranged exactly as the regression equation (10.1) with the same coefficients as obtained by the traditional multi-linear least-square procedure [12].

However, apart in delivering an alternative (analytical) way of structure-activity relationships the algebraic spectral picture offers the possibility of introducing a new phenomenology in correlation analysis. Actually, since the generalized scalar product of two N-dim vectors is considered,

$$\langle \Psi_l | \Psi_k \rangle = \sum_{i=1}^{N} \psi_{il}\psi_{ik} = \langle \Psi_k | \Psi_l \rangle, \qquad (10.2)$$

the associated norms of observed/predicted activity/model may be as well introduced,

$$\left\| Y \rangle_{PREDICTED}^{OBSERVED/} \right\| = \sqrt{\sum_{i=1}^{N} (y_i^2)_{PREDICTED}^{OBSERVED/}}, \qquad (10.3)$$

allowing for new algebraic further definition of the correlation index of a given model as the ration between the computed and measured activity norms [12a]:

$$r^{ALG} = \frac{\left\| |Y\rangle^{PREDICTED} \right\|}{\left\| |Y\rangle^{OBSERVED} \right\|}.$$

(10.4)

This new correlation index, apart of being of pure algebraic nature, has many phenomenological advantages in terms of mechanistic QSAR interpretation and prediction. Firstly, being based on scalar norm, while viewing it as the "length" of the observed/predicted (eco)biological action (the so called *second SPECTRAL-SAR ecotoxicological principle* [12c]), the algebraic correlation unveils as the "intensity" of the chemical-(eco)-bio-interaction determined by the ratio of the expected to measured activity lengths (the so called *third SPECTRAL-SAR ecotoxicological principle* [12c]). There was recently proved that this intensity always beholds to (0, 1) realm with a systematic superiority over the consecrated statistical correlation factor based on squared sum of residues and on statistical variance of the predicted endpoints [12e]. This way, the algebraic correlation index measures how close the length of a certain QSAR model approaches the measured length of the studied bioactivity. Now, from "length" to the "path" idea there is just a step away in QSAR analysis.

Since, each computed model may be now characterized by the specific predicted activity norm and algebraic correlation index, one can consider the paths across various computed models in *spectral norm-algebraic correlation* hyperspace, based on the elementary Euclidian measure:

$$[A, B] = \sqrt{\left(\left\| |Y^B\rangle \right\| - \left\| |Y^A\rangle \right\| \right)^2 + \left(r_B^{ALG} - r_A^{ALG} \right)^2}, \text{A, B—endpoints.}$$

(10.5)

Next, once having a path-operational definition the "selection" of optimum model may be performed throughout imposing the so called "ergodic" least-action algebraic path principle

$$\delta \left[A_{predicted}^{uni-parameter}, A_{predicted}^{b-parameter} \ldots A_{predicted}^{M-parameter}, A_{observed} \right] = 0$$

(10.6)

such that, starting from each considered uni-parameter model all bi-, many-, until the *M*-parameter endpoints be included along the paths connecting them, successively, in all possible unique ways approaching observed activity.

With this condition the "ergodic" endpoint selection principle was advanced in algebraic manner allowing the identification of the uni-parameter models (i.e., the specific structural parameter) responsible for "the first cause" in producing an output action if it originates on the shortest (the *alpha*) path computed.

If two paths originating in different uni-parameter models are equal there will be selected that one with the closest uni-parameter model norm-length to the observed activity one. If two paths originating in the same model are equal, then the selected path will be that one providing the shortest "first move", that is, that path containing the bi-parameter model having the closest norm-length to the uni-parameter starting model will be selected. There is obvious that the least-path principle is iteratively applied equally at global level linking uni-parameter models to observed activity endpoint as well at the local inter-endpoints level. Naturally, there will be generated so many paths as many structural parameters are considered; their hierarchy will finally provide the mechanistic picture of the structural causes that determine (through primary path) and influence (through the secondary, etc., paths) the observed bioactivities.

Remarkably, within the SPECTRAL-SAR approach, the validation and prediction issues are resolved by the models contained by the selected alpha path. This way, the algebraic spectral approach offers, in principle, a complete picture of both mechanistic and prediction facets of a QSAR analysis, without relaying on statistical tools and assumptions. However, the reliability of algebraic phenomenology has to be confronted with the traditional statistical tests both at conceptual and computation levels of comprehension. An illustrative example is in next section detailed.

ALIPHATIC AMINES' TOXICITY AND THEIR SPECTRAL-SAR ANALYSIS

Amines are organic compounds and functional groups that contain a basic nitrogen atom with a lone pair. Amines are derivatives of ammonia, wherein one or more hydrogen atoms are replaced by organic substituents such as alkyl and aryl groups [14].

Primary amines arise when one of three hydrogen atoms in ammonia is replaced by an organic substituent. *Secondary amines* have two organic substituents bound to N together with one H. In *tertiary amines* all three hydrogen atoms are replaced by organic substituents. It is also possible to have four alkyl substituents on the nitrogen. These compounds have a charged nitrogen center, and necessarily come with a negative counterion, so they are called quaternary ammonium salts. The functional group within the primary amines is the amino-group. The secondary amines are distinguished by the secondary amino-group, which means that one hydrogen was replaced by another substituent. In the case of tertiary amines a tertiary nitrogen atom is bond to three substituents: $R-NH_2$ (primary amines), R_2NH (secondary amines), R_3N (tertiary amines).

In the geosphere may enter the primary amines (fatty) because of their use as flotation agents and reach the soil directly. Total emissions of fatty amines into the hydrosphere from the manufacturing and from chemical processing (excluding salt formation) amounted to less than 200 kg per year, or, in respect to the individual fatty amines, less than 30 kg for each [15]. Primary fatty amines are interface-active compounds due to their ability to form protons simply and this property explains the strong adsorption potential. Thus, a high geoaccumulation potential due to their physi- and chemisorption on inorganic soil components can be expected [15]. Due to their ability to form protons simply, the primary fatty amines are interface-active compounds, this property explains the strong adsorption potential. Thus, a high geoaccumulation potential because of their physi- and chemisorption on inorganic soil components can be expected [15].

Some of the most important aliphatic amines, along their (eco)toxicity characterization follow below.

Methylamine is the chemical compound with a formula of CH_3NH_2. It is a deriva-tive of ammonia, where in one H atom is replaced by a methyl group. It is the simplest primary amine. It is usually sold as solutions in methanol (2M), ethanol (8M), and water (40%), or as the anhydrous gas in pressurized metal containers. It has a strong odor similar to rotten fish. Methylamine is used as a building block for the synthesis of other organic compounds, including many illicit drugs; in the United States, the Drug Enforcement Administration (DEA) lists methylamine as a precursor, and purchases of any significant quantity are likely to arouse law enforcement attention [16, 17].

Dimethylamine is an organic compound with the formula $(CH_3)_2NH$. This second-ary amine is a colorless, flammable liquid with an ammonia- or fish-like odor. Dimeth-ylamine is generally encountered as a solution in water at concentrations up to around 40%. In 2005, an estimated 270,000 tonnes were produced [18]. Dimethylamine reacts with acids to form salts, such as dimethylamine hydrochloride, an odorless white solid with a melting point of 171.5°C. Dimethylamine is produced by catalytic reaction of methanol and ammonia at elevated temperatures and high pressure [16]. Dimethyl-amine is a precursor to several industrially significant compounds [18]. It reacts with carbon disulfide to give dimethyldithiocarbamate, a precursor to a family of chemi-cals widely used in the vulcanization of rubber. It is raw material for the production of many agrichemicals and pharmaceuticals, such as dimefox and diphenhydramine, respectively. The chemical weapon tabun is derived from dimethylamine. dimethyl-amine is utilizes like a pheromone for communication [19].

Piperidine is an organic compound with the molecular formula $(CH_2)_5NH$. This heterocyclic amine consists of a six-membered ring containing five methylene units and one nitrogen atom. It is a colorless fuming liquid with an odor described as am-moniacal, pepper-like. Piperidine is a widely used building block in the synthesis of organic compounds, including pharmaceuticals. The piperidine structural motif is present in numerous natural alkaloids such as quinine and piperine, the main ac-tive chemical agent in black pepper and relatives (*Piper sp.*—the family Piperaceae), hence the name. A significant industrial application of piperidine is for the production of dipiperidinyl dithiuram tetrasulfide, which is used as a rubber vulcanization accel-erator [20]. Otherwise piperidine and its derivatives are widely used building blocks in the synthesis of pharmaceuticals. The piperidine structure is found for example in the following drugs: raloxifene, minoxidil, thioridazine, haloperidol, droperidol, and mesoridazine. Piperidine is listed as a precursor under the United Nations Convention Against Illicit Traffic in Narcotic Drugs and Psychotropic Substances due to its use (peaking in the 1970s) in the clandestine manufacture of PCP (also known as angel dust), see ref. [21]—"List of precursors and chemicals frequently used in the illicit manufacture of narcotic drugs and psychotropic substances under international con-trol." Piperidine, also a secondary aliphatic amine, is a member of a group of biogenic amines and is present in the central nervous system of both invertebrate and vertebrate [17]. It is a volatile base with nicotine-like action and was regarded as endogenous "synaptotropic substance" (von Euler, 1945; as cited in Giacobini, 1976) [17].

Trimethylamine is an organic compound with the formula $N(CH_3)_3$. This colorless, hygroscopic, and flammable tertiary amine has a strong "fishy" odor in low concentrations and an ammonia-like odor at higher concentrations. It is a gas at room temperature but is usually sold in pressurized gas cylinders or as a 40% solution in water. Trimethylamine is a product of decomposition of plants and animals. It is the substance mainly responsible for the odor often associated with fouling fish, some infections, and bad breath. It is also associated with taking large doses of choline and carnitine. Trimethylamine is used in the synthesis of choline, tetramethylammonium hydroxide, plant growth regulators, strongly basic anion exchange resins, and dye leveling agents. Gas sensors to test for fish freshness detect trimethylamine [22]. Trimethylaminuria, also known colloquially as the fish malodor syndrome, provides an excellent example of how genetically determined variability in the metabolism of a dietary derived chemical, namely trimethylamine, can result in a distressing clinical condition. At a more general level it reflects just how genetic constitution can adversely influence interactions with one's diet. In the past 15 years much has been learned about this particular metabolic disorder, at one time considered to be extremely rare in its occurrence, but in the light of new information, opinions about this now have to be revised [23].

Tributylamine, a tertiary aliphatic amine, is used as intermediate, as a catalyst and acid acceptor in organic synthesis and polymerization, as corrosion inhibitor and as solvent and auxiliary for isolating and purifying antibiotics [24]. Inhalation of vapors of tributylamine irritates the respiratory tract. May cause distressed breathing, coughing. It may cause central nervous system effects. At room temperature, the substance has such a low vapor pressure that inhalation of the vapor is unlikely. Discharge into the environment is estimated to be less than 5 kg per year in Germany and occurs almost via waste water. No information about the occurrence in the atmosphere or in soil and sediments were available. In the 1970s tributylamine was found in the river Rhine in a concentration range of 0.1–1 μgL^{-1} [24].

Dimethylethylamine, also a tertiary aliphatic amine, is used as a catalyst in the polymerization of polyurethane [25]. It is a highly water-soluble and volatile compound and its concentration in the ambient air may be high [26]. Amines and their derivatives are compounds which are extensively used as drugs, cosmetics, dyes, in synthesis of pesticides, and as synthetic intermediates. Aliphatic amines are considered to be strong organic bases [27]. The production of primary fatty amines (in 1992) with a chain length of C_8 to C_{18} was about 8000 tonnes and, for example the production of octylamine, was like 450–500 tonnes [15]. About 75% of the primary fatty amines are used as intermediates for synthesizing ethoxylated fatty amines. These are widely used as cationic, surface-active substances, for example in auxiliary agents for dyeing and textile, as additives for mineral oil and as antistatic agents for plastics. About 25% were directly used in form of their salts, mainly as flotation agents, as dispersing agents for pigments or as corrosion inhibitors [15]. In the United States of America 2,000–3,000 tonnes of cyclohexylamine are used each year as corrosion inhibitor in steam lines and boiler heating systems. Cyclohexylamine is suspected to have a teratogenic, mutagenic, and carcinogenic potential [28]. A mammalian fertility study suggests that cyclohexylamine targets Sertoli cells in the testes [29].

Morpholine is an organic chemical compound having the chemical formula $O(CH_2CH_2)_{2NH}$. This heterocycle, pictured at right, features both amine and ether functional groups. Because of the amine, morpholine is a base; its conjugate acid is called morpholinium. For example, when morpholine is neutralized by hydrochloric acid, one obtains the salt morpholinium chloride. Morpholine is a common additive, in ppm concentrations, for pH adjustment in both fossil fuel and nuclear power plant steam systems [30]. Morpholine is used because its volatility is about the same as water, so once it is added to the water, its concentration becomes distributed rather evenly in both the water and steam phases. Its pH adjusting qualities then become distributed throughout the steam plant to provide corrosion protection. Morpholine is often used in conjunction with low concentrations of hydrazine or ammonia to provide a comprehensive all-volatile treatment chemistry for corrosion protection for the steam systems of such plants. Morpholine decomposes reasonably slowly in the absence of oxygen even at the high temperatures and pressures in these steam systems [30]. The production of morpholine, was about 12,000 tonnes in 1988 in Germany and more than 75% is exported [31]. The amounts entering the atmosphere during manufacture are likely to be small. However, the entry of morpholine into the environment is accompanied by non-quantifiable amount of the nitrosation product, N-nitrosomorpholine [31]. Nitrosation products are known to have a carcinogenic potential. Morpholine is widely used in organic synthesis. For example, it is a building block in the preparation of the antibiotic linezolid and the anticancer agent gefitinib (Iressa). Morpholine is used as a chemical emulsifier in the process of waxing fruit. Fruits make waxes naturally to protect against insects and fungal contamination, but this can be lost by means of the food processing companies when they clean the fruit. As a result, an extremely small amount of new wax is applied and morpholine is then added and used as an emulsifier to evenly coat a fruit with the wax. In research and in industry, the low cost and polarity of morpholine lead to its common use as a solvent for chemical reactions.

Through amines and their derivatives are compounds which are extensively used as drugs, cosmetics, dyes, in synthesis of pesticides, and as synthetic intermediates, since considered to be strong organic bases [32], a broad classification scheme was developed by Verhaar and co-workers in 1992 according which compounds are assigned to one of four general classes of toxic mode of action corresponding to short-term exposures with lethality as endpoint: inert chemicals, less inert chemicals, reactive chemicals, and specific-acting chemicals [33].

Primary aliphatic amines are slightly more toxic than baseline toxicity and were classified as less inert compounds acting by a "polar narcosis" mechanism. In general, narcosis in aquatic organisms is described as a non-specific reversible functional disturbance of biological membranes due to the accumulation of chemicals in hydrophobic phases within the organism. Instead, secondary and tertiary aliphatic amines were classified as inert compounds (displaying narcotics or baseline toxicity), because of their non-specific mode of action with a general coherency between lipophylicity and toxicity was suggested [34].

The biotransformation of amines is catalyzed by the cytochrom P 450-system, an important system for the metabolism of foreign compounds which is located in the endoplasmatic reticulum, in the membranes of mitochondria and in plasmatic membranes of prokaryotes. The N-oxidation is the most important metabolic way for N-containing chemicals in humans and animals [35]; as such:

- primary amines can be oxidized to hydroxylamines and then to oximes:

$$R\text{-}CH_2\text{-}NH_2 \xrightarrow{+\,[O]} R\text{-}CH_2\text{-}NH\text{-}OH \longrightarrow R\text{-}CH=N\text{-}OH$$

- secondary amines can be oxidized to hydroxylamines and then, if a hydrogen atom is available at the α-C atom, to nitrones:

$$R''\!-\!\overset{\overset{H}{|}}{\underset{\underset{R'}{|}}{C}}\!-\!\overset{\overset{}{}}{\underset{\underset{H}{|}}{N}}\!-\!R \xrightarrow{+\,[O]} R''\!-\!\overset{\overset{H}{|}}{\underset{\underset{R'}{|}}{C}}\!-\!\overset{\overset{OH}{|}}{N}\!-\!R \longrightarrow R''\!-\!\overset{}{\underset{\underset{R'}{|}}{C}}\!=\!\overset{+}{N}\!-\!R \rightleftharpoons R''\!-\!\overset{}{\underset{\underset{R'}{|}}{\overset{+}{C}}}\!-\!\overset{\overline{|\overline{O}|}^{-}}{N}\!-\!R$$

- tertiary amines can be transformed by microsomal enzymes to *N*-oxide-metabolites which then can be further desalkylated and/or reduced:

$$R''\!-\!\overset{}{\underset{\underset{R'}{|}}{N}}\!-\!R \xrightarrow{+\,[O]} R''\!-\!\overset{\overset{|\overline{O}|^{-}}{|}}{\underset{\underset{R'}{|}}{\overset{+}{N}}}\!-\!R$$

The most important property of aliphatic amines is their basic character. If an aliphatic amine is dissolved in water the pH will increase due to the protonation and alkylammonium ions and hydroxide ions will be formed. Like ammonia, most amines are Brönsted and Lewis bases, but their base strength can be changed enormously by substituents. Most simple alkyl amines have pK_a's in the range 9.5–11.0, and their water solutions are basic (have a pH of 11–12, depending on concentration).

Theoretical scheme [9, 12, 23] of the dissociation of amines in water:

$$R\!-\!NH_2 + HOH \rightleftharpoons R\!-\!\overset{+}{N}H_3 + HO^{-}$$

Amines have found uses in an amazing array of applications. Following is a partial list:

- *Agrochemical*: Alkylamines are important chemical intermediates used in the synthesis or formulation of many types of herbicides and pesticides.
- *Monomers, Polymers, and Rubber*: Amines act as intermediates for a wide variety of monomers and polymers for applications as diverse as resins, plastics, rubber, and textiles.
- *Petroleum and Petrochemical*: Emulsion breakers, antioxidants, sludge prevention, lubricating oil additives, and gas treating.
- *Corrosion Inhibition*: Metalworking fluids, lubricants, and coatings. Corrosion inhibitors which contain nitrogen, obtained by an amine, are used in many industries.
- *Neutralizing Agent*: For pharma synthesis, textile fibers, water treatment, and refinery gas treatment.
- *Curing Agents*: For epoxy or polyurethane resins.
- *Amines Salt Providers*: For agrochemical formulations, anti-corrosion additives for water treatment or lubricants.

- *Building Block for Fine Chemical Synthesis*: Active ingredients for pharmaceutical, agrochemistry, or specialty amines manufacturing.

Amines are used in a wide variety of industries:

- Pharmaceuticals
- Agrochemicals
- Paints and coatings
- Specialty chemicals (plastic additives, monomer synthesis, water treatment, etc.)

Short chain alkyl amines are used as raw materials of solvent, alkyl alkanolamines, and ingredients of rocket fuels. They are used to make other organic chemicals including rubber vulcanization accelerators, pesticides, quaternary ammonium compounds, photographic chemicals, corrosion inhibitors, explosives, dyes, and pharmaceuticals; are used, also, in rayon and nylon industry to improve the tensile strength. Alkylamines are used as intermediates for ion exchange resins, water soluble polymers, pharmaceuticals, herbicide softeners, polymerization initiators, rubber chemicals, and cross-linking agents. Amines are used as reducing agents for the recovery of precious metals. They are versatile intermediates. They have active applications in organic synthesis for polymerization catalyst, chain extender in urethane coatings, agrochemicals, pharmaceuticals, photographic, heat stabilizers, polymerization catalysts, flame-retardants, blowing agents for plastics, explosives, and colorants. Long chain alkyl amines are used for the synthesis of organic chemicals and surfactants used as a corrosion inhibitor, detergent, ore floating agent, fabric softener, anti-static agent, germicide, insecticide, emulsifier, dispersant, anti-caking agent, lubricant, and water treatment agent. Alkyl tertiary amines are used as fuel additives; they have similar applications with long chain alkyl amines. Hexamethylenediamine used in the manufacture of nylon-6,6 is prepared by catalytic addition of hydrogen to nitriles [16].

Given the multiple uses of amines in a wide variety of environmental and industrial fields, the present interest in QSAR modeling of their toxicity among primary, secondary, and tertiary structures stands on the forefront importance for environment and life monitoring and control.

The rat oral 50% lethal dose (LD_{50}) toxicity is here considered for the amines of Table 10.1. The envisaged structure-activity analysis will be performed according with the Hansch description,

$$A = b_0 + b_1 \left(\frac{hydrophobic}{descriptor} \right) + b_2 \left(\frac{electronic}{descriptor} \right) + b_3 \left(\frac{steric}{descriptor} \right), \tag{10.7}$$

throughout correlating the genuine (non-biased) biotoxicity $A = Log_{10}(1/LD_{50})-6$ with molecular parameters as hydrophobicity LogP, electronic polarizability POL, and steric total energy E_{tot} computed at the optimum geometric molecular configuration evaluated within the HyperChem environment [36]. Since the usual statistic analysis demands the *trial* and *test* stages in validation the molecules of Table 10.1 were classified accordingly based on the best fulfillment of the normal distribution of input data (LD50), as evidenced from Figure 10.1, however, such that each category of aliphatic amines to be represented in both trial and test sets of toxicants. The computed end-points are uni-, bi-, and three-parameters containing models with regression equations together with statistic and algebraic results collected in the Table 10.2. The basic statistic descriptors and tests employed are built on the observed, predicted, and residues (errors) squared sums of differences:

$$SO = \frac{1}{N-1}\sum_{i=1}^{N}\left(A_i^{obs} - \overline{A}\right)^2, \; SP = \frac{1}{M}\sum_{i=1}^{N}\left(A_i^{pred} - \overline{A}\right)^2, \; SE = \frac{1}{N-M-1}\sum_{i=1}^{N}\left(A_i^{obs} - A_i^{pred}\right)^2, \quad (10.8)$$

with $\overline{A} = N^{-1}\sum_{i=1}^{N} A_i^{obs}$. They are traditionally evaluated and validated as [2]:

- Standard error of estimates is introduced as

$$SEE = \sqrt{SE}, \quad (10.9a)$$

while its relevance to the goodness of the fit is tested by comparing it with the max–min range of observed activity throughout the index ratio

$$\rho_{SEE} = \left|A_{max}^{obs} - A_{min}^{obs}\right| / \sqrt{SE} > 5; \quad (10.9b)$$

- Explained variance is defined with the form:

$$EV = R_{adjusted}^2 = 1 - \frac{SE}{SO} \quad (10.10)$$

from where the simple statistical correlation factor R follows by skipping out the degree of freedom dependence in (10.8) for the involved sums in (10.10);

- The Student-t index and test are considered as:

$$t_{computed \atop (N-M-2)} = \sqrt{N-1-(M+1)}\sqrt{\frac{EV}{1-EV}} > t_{tabulated \atop (1-\alpha=0.995, \atop N-M-2)} = \begin{cases} 3.169, & N=15, M=3 \\ 3.106, & N=15, M=2 \\ 3.055, & N=15, M=1 \end{cases} \quad (10.11a)$$

while the degree of fulfilling of this criteria is indicated by the ration index rule:

$$\rho_{t-comp} = t_{computed \atop (N-M-2)} / t_{tabulated \atop (1-\alpha=0.995, \atop N-M-2)} > 2 \; ; \quad (10.11b)$$

- The Fisher index and test are considered as:

$$F_{computed \atop (M,N-M-1)} = \frac{SP}{SE} > F_{tabulated \atop (1-\alpha=0.99; \atop M,N-M-1)} = \begin{cases} 6.22, & N=15, M=3 \\ 6.93, & N=15, M=2 \\ 9.07, & N=15, M=1 \end{cases} \quad (10.12a)$$

while the degree of fulfilling of this criteria is indicated by the ration index rule:

$$\rho_{F-comp} = F_{computed \atop (M,N-M-1)} / F_{tabulated \atop (1-\alpha=0.9 \; ; \atop M,N-M-1)} > 4 \quad . \quad (10.12b)$$

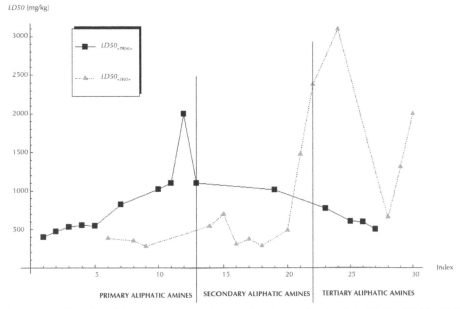

Figure 10.1. The plot of the primary, secondary, and tertiary amines' toxicity LD50 of Table 10.1 grouped within trial and test sets.

Table 10.1. The series of primary, secondary, and tertiary aliphatic amides together with associated 50% lethal toxicity dose LD_{50} (mg/kg) [13a] (when original data was given as an interval the average range was here considered) among the employed genuine (non-biased) activity $A = Log_{10}(1/LD_{50}) - 6$ and structural parameters as hydrophobicity LogP, polarizability POL (E^3) and total energy at optimal molecular configuration E_{tot} (kcal/mol) computed within Hyperchem environment [36]. The molecules included in the "trial" set have their index marked with a filled square while those considered in the "test" set were marked with the filled grey triangle, see also the Figure 10.1.

Index	CAS No. CAS Name	Structural Formula	LD_{50}	A	LogP	POL	E_{tot}
		▼ **PRIMARY ALIPHATIC AMINES** ▼					
1■	75-04-7 Ethylamine	$H_3C\text{-}CH_2\text{-}NH_2$	400	−602	−0.27	5.80	−12912.76
2 ■	107-10-8 Propylamine	$H_3C\text{-}CH_2\text{-}CH_2\text{-}NH_2$	470	−2.672	0.20	7.63	−16506.47
3 ■	109-73-9 Butylamine	$H_3C\text{-}CH_2\text{-}CH_2\text{-}CH_2\text{-}NH_2$	528	−2.722	0.59	9.47	−20100.16
4 ■	75-31-0 Isopropylamine	$H_3C\text{-}CH\text{-}CH_2\text{-}CH_3$ NH_2	550	−2.740	0.14	7.63	−16505.39
5 ■	13952-84-6 Sec-Butylamine	$H_3C\text{-}CH\text{-}CH_2\text{-}CH_3$ NH_2	545	−2.736	0.61	9.47	−20097.96

Table 10.1. *(Continued)*

Index	CAS No. CAS Name	Structural Formula	LD$_{50}$	A	LogP	POL	E$_{tot}$
6 ▲	108-91-8 Cyclohexylamine	⬡–NH$_2$	385	–2.585	0.97	12.36	–26650.52
7 ■	2869-34-3 1-Tridecanamine	C$_{13}$H$_{27}$-NH$_2$	820	–2.913	4.16	25.98	–52443.68
8 ▲	111-86-4 1-Octanamine	CH$_3$-(CH$_2$)$_7$-NH$_2$	350	–2.544	2.18	16.80	–34475.05
9 ▲	2016-57-1 1-Decanamine	CH$_3$-(CH$_2$)$_9$-NH$_2$	280	–2.447	2.97	20.47	–41662.51
10 ■	124-22-1 Dodecylamine	CH$_3$-(CH$_2$)$_{11}$-NH$_2$	1020	–3.008	3.77	24.14	–48849.95
11 ■	2016-42-4 tetradecylamine	CH$_3$-(CH$_2$)$_{13}$-NH$_2$	1100	–3.041	4.56	27.81	–56037.41
12 ■	124-30-1 Octadecylamine	CH$_3$-(CH$_2$)$_{17}$-NH$_2$	2000	–3.301	6.14	35.15	–70412.31
13 ■	112-90-3 Oleyl amine	CH$_3$-(CH$_2$)$_7$-CH = CH-(CH$_2$)$_7$-CH$_2$-NH$_2$	1100	–3.041	5.88	34.96	–69753.33
		▼SECONDARY ALIPHATIC AMINES▼					
14 ▲	109-89-7 Diethylamine	CH$_3$-CH$_2$-CH$_2$-NH-CH$_2$-CH$_2$-CH$_3$	540	–2.732	0.48	9.47	–20090.93
15 ▲	142-84-7 Dipropylamine	CH$_3$-CH$_2$-CH$_2$-NH-CH$_2$-CH$_2$-CH$_3$	695	–2.841	1.42	13.14	–27278.29
16 ▲	13360-63-9 Butylamine-N-ethyl	CH$_3$-CH$_2$-CH$_2$-CH$_2$-NH-CH$_2$-CH$_3$	308.5	–2.489	1.35	13.14	–27278.29
17 ▲	111-92-2 Dibutylamine	CH$_3$-CH$_2$-CH$_2$-CH$_2$-NH-CH$_2$-CH$_2$-CH$_2$-CH$_3$	369.5	–2.567	2.21	16.80	–34465.66
18 ▲	101-83-7 Dicyclohexylamine	(dicyclohexylamine structure) NH	286.5	–2.457	2.97	22.60	–47562.52
19 ■	106-20-7 Dihexylamine- 2,2-diethyl	(dihexylamine-2,2-diethyl structure) NH	1015	–3.006	5.39	31.48	–63205.53
20 ▲	110-89-4 Piperidine	(piperidine structure) NH	485	–2.686	0.52	10.53	–23051.82

Table 10.1. *(Continued)*

Index	CAS No. CAS Name	Structural Formula	LD_{50}	A	LogP	POL	E_{tot}
21 ▲	110-91-8 Morpholine		1475	−3.169	−0.55	9.33	−26841.67
22 ▲	6485-55-8 cis-2,6-dimetyl- morpholine		2380	−3.377	0.28	13.00	−34022.86
	▼TERTIARY ALIPHATIC AMINES▼						
23 ■	75-50-3 Trimethylamine		770	−2.886	0.16	7.63	−16487.84
24 ▲	593-81-7 Trimethylamine- hydrochloride		3090	−3.49	0.07	9.37	−25441.78
25 ■	598-56-1 N,N – dimethyl- ethyl- amine		606	−2.782	0.50	9.47	−20079.71
26 ■	121-44-8 Triethylamine		595	−2.775	1.18	13.14	−27262.98
27 ■	98-94-2 Cyclohexylamine- N,N –dimethyl		499	−2.698	1.74	16.03	−33812.66
28 ▲	102-82-9 Tributylamine		654	−2.816	3.78	24.14	−48824.4
29 ▲	112-18-5 Dodecylamine- N,N- dimethylamine		1315	−3.119	4.53	27.81	−56016.79
30 ▲	4088-22-6 1-Octadecanamine- N-methyl-N-octa- decyl		2000	−3.301	13.67	70.02	−138667.22

Table 10.2. Structure-activity relationships for all possible correlation models considered from the data in Table 10.1 together with the statistical (simple correlation factor, standard error of estimation SEE, explained variance EV, Student t-test, Fischer F-test) and algebraic (correlation factor r^{ALG} and norm-length $\|\bullet\|$) descriptors for each considered endpoint.

Mode	End point	Structure-activity Relationships	Statistical descriptors								Algebraic descriptors	
			$R^{\spadesuit}$	SEE^{*}	ρ_{SEE}^{*}	$EV^{¥}$	$t_{comp}^{§}$	$\rho_{t\text{-}comp}^{§}$	$F_{comp}^{\#}$	$\rho_{F\text{-}comp}^{\#}$	$\left\|Y^{predicted}_{MODE}\right\|^{\bullet}$	$r^{ALG\spadesuit}$
Ia	LogP	−26981 −0.0705451 LogP	0.887662	0.09004	7.764	0.772	6.3676	2.08432	48.305	5.3258	11.1003228	0.999573
Ib	POL	−2.59298 −0.0151562 POL	0.880475	0.09269	7.541	0.758	6.1299	2.00651	44.839	4.94366	11.1000384	0.999547
Ic	E_{tot}	−2.58076 +7.73519·10⁻⁶ E_{tot}	0.878731	0.09332	7.490	0.755	6.0752	1.98861	44.06	4.85777	11.0999697	0.999541
IIa	LogP, POL	−3.16218 −0.37495 LogP +0.0659889 POL	0.901464	0.08809	7.936	0.781	6.2708	2.01893	26.023	3.75512	11.1008756	0.999622
IIb	LogP, E_{tot}	−3.19725 −0.362757 LogP −3.2401510⁻⁵ E_{tot}	0.904036	0.08699	8.036	0.787	6.3719	2.05148	26.837	3.87258	11.1009795	0.999632
IIc	POL, E_{tot}	−2.79473 −0.25427 POL −1.2228510⁻⁴ E_{tot}	0.893712	0.0913	7.656	0.765	5.9869	1.92753	23.809	3.43564	11.1005641	0.999594
III	LogP, POL, E_{tot}	−3.17409 −0.304737 LogP −0.0896505 POL +7.18245·10⁻⁵ E_{tot}	0.905222	0.09032	7.739	0.771	5.789	1.82676	16.6391	2.6751	11.1010275	0.999636

$\spadesuit$ see equation (10.10) and the following note; * see equation (10.9); ¥ see equation (10.10); § see equation (10.11); # see equation (10.12); $\bullet$ see equation (10.3) with $\|Y^{observed}\|$ =11.1051; $\spadesuit$ see equation (10.4).

From Table 10.2 one gets, indeed, a somehow confusing situation since, according with individual criteria of statistical validation, there resulted that each group of statistical indices provides different model as the most robust one among all trials. Actually, based only on direct comparison of simple statistical correlation factors the model LogP + POL + E_{tot} (III) is predicted as the most reliable one from the results of Table 10.2. Instead, when analyzed throughout the minimum of standard error of estimation (SEE), and maximum of its associate observed activity range normalized index, as well as by the maximum of explained variance (EV) across all investigated models of Table 10.2, the LogP + E_{tot} model (IIb) is selected as the recommended regression.

Moreover, when the Student-t and Fisher tests respecting the degree with which their computed values exceed the corresponding tabulate ones, see equations (10.11) and (10.12), the model LogP (Ia) it is yielded as the optimum one out from the same set of models investigated. We may equally called this situation as "statistical paradox" in which, having a collection of models, depending on the statistical tool used different model will be assessed as the most robust one. Note that the values of algebraic correlations close to 1.00 in Table 10.2 give already the sign of the above mentioned statistical paradox.

Fortunately, having introduced the idea of paths across models such paradox is somehow solved out if the different "successful" models are arranged in a linked manner along a "statistical path," here with the form

$$\text{Ia} \rightarrow \text{IIb} \rightarrow \text{III.} \tag{10.13}$$

This would be the end of the statistical analysis of a QSAR study, since from such path also the mechanistic interpretation of the bio(eco)logical-chemical-action may be formulated as originating in hydrophobic cause (LogP from model Ia) followed by entering "in action" of the stericity (through E_{tot} from model IIb over the LogP presence—already included in the model Ia) toward the final stabilization by the ionic or electrostatic character of the interaction (by POL of model III in addition to LogP and E_{tot} from the previous ones).

Further on, for processing the algebraic information of concerned endpoints, the specific norms and algebraic correlation factors are presented in Table 10.2 for each of studied models; as a note, in all cases the algebraic factor is clearly dominating its statistical counterpart being, however, so high that many more digits have to be considered for making them different; this is nothing than the reflection of the very specific (or local) character of the modeled interaction. In other words, there is suggested that, among all similar models produced on the same base and algorithm applied on a given data, the difference in their reliability and prognosis is as sharp as the degree of interaction characterized.

Making a step forward, with the algebraic information of Table 10.2 all the possible (ergodic) paths are constructed and their lengths computed according with the recipe of equation (10.5) with results displayed in the Table 10.3.

Table 10.3. Synopsis of the algebraic paths connecting the SPECTRAL-SAR models of Table 10.2 in the algebraic norm-correlation hyperspace; the statistical path emerging from analysis of statistical descriptors of Table 10.2 is also marked.

Algebraic Paths	Value	
Ia-IIa-III	0.00070756	
Ia-IIb-III	0.00070756	*Statistical-path*
Ia-IIc-III	**0.00070756**	*Algebraic-α path*
Ib-IIa-III	0.000993177	*Algebraic-β path*
Ib-IIb-III	0.000993177	
Ib-IIc-III	0.000993177	
Ic-IIa-III	0.00106214	
Ic-IIb-III	0.00106214	*Algebraic-γ path*
Ic-IIc-III	0.00106214	

The algebraic SPECTRAL-SAR analysis is completed with the search of the minimum path out from those listed in Table 10.3 by means of the least action principle of equation (10.6). There is noted that all paths originating on the same uni-parameter models produce equal lengths toward the observed action (in fact, toward the computed endpoint III since the path from III to observed endpoint is always the same in any uni- and bi-parameter linked models); in this situation the discrimination among them is made on above described global-to-local minimum principle. Practically, one chooses the global minimum path among all existing path lengths; here results that there are three such paths, all starting from Ia model; then, the local differentiation is made by looking to the closest algebraic norm among IIa, IIb, and IIc models respecting the norm of Ia endpoint: it results that the model IIc displays in Table 10.2 such feature so that it is the most susceptible to be considered as the "first move" out from the model Ia. This way the "algebraic path"

$$Ia \rightarrow IIc \rightarrow III \tag{10.14}$$

was selected by the algebraic spectral methodology as the primary or alpha path in linking the causes relaying on considered structural parameters toward the observed effects as measured ecotoxicity. In the same manner, the other two influential (beta and gamma) paths are determined from the remaining basin of inter-endpoints pathways, based on the same systematic principle, while excluding the presence of previous selected models in the actual search (i.e., if the model IIb was already included in the alpha path it will be excluded from the search of beta path and so on).

Now, we need to discuss about the two above selected paths, namely the statistical and algebraic ones, equations (10.13) and (10.14), respectively.

First, about their noted similitude, they both belong to the same algebraic path length, as clearly visualized from Table 10.3; this nevertheless, emphasizes on the power of least action principle in selecting both the statistical and algebraic path as belonging to the global minimum paths' hierarchy; note that in the previous exposed "algebraic language" the same length means the same "intensity of interaction" that is equally provided by statistical and algebraic-spectral approaches. Also observe that without the least action principle and of its present particularizations in Table 10.3, the statistical results obtained from Table 10.2 would remain without a consistent rationale; now, not only the "statistical path" is seen as one of the possible paths in connecting the possible endpoints for the given set of structural parameters but it is implicitly recovered among the shortest paths by the least action principle; therefore, the statistical results of Table 10.2 are validated within the present algebraic-spectral frame of QSAR analysis.

Yet, it remains to distinguish between the statistical and algebraic paths in order to establish which of them is more reliable in describing the molecular mechanisms of action in producing the observed bioeffects. For deciding upon this issue the test set of data of Table 10.1 is now employed to all involved models in paths of equations (10.13) and (10.14) to produce the predicted-test activities to be compared with those observed in Table 10.1. The results are depicted in Figure 10.2; at the first sight they are of poor quality but this is due to the non-normal distribution of selected test-data as clearly evidenced from Figure 10.1. Still, because of their dispersion such data may be used for ordering the predictability power among our trial-selected models. From Figure 10.2 one can distinguish that the model IIc has not only smooth supremacy over its "concurrent" model IIb but also displays better predictability power over all other models involved, Ia and III. Therefore, we may conclude that the alpha algebraic path of equation (10.14) is the optimum one both by minimum length among the endpoint parameters involved over the trial molecules and for the test compounds of Table 10.1.

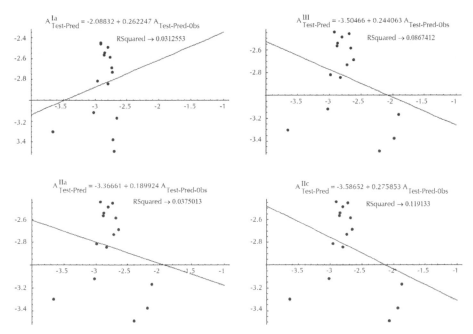

Figure 10.2. Comparative observed-predicted plots and their regression performances for the test data of Table 10.1 employed with the models Ia and III (upper left and right) and IIb and IIc (bottom left and right) of the Table 10.2, respectively.

The resulted optimum endpoint in trial-tested path among the structural parameters LogP, POL, and E_{tot} computed for the sets of aliphatic molecules of Table 10.1 in modeling of their (anti-activity of) LD50 toxicity on rats stands therefore the path of equation (10.13); it prescribes that the superposition of the structural (quantum) molecular states characterized by $|LogP\rangle$ (model Ia based) and $|POL, E_{tot}\rangle$ (model IIc based) determine throughout the resultant model III the observed effects in toxicity. Such as, the spectral-selected path gives information not only on the order of structural parameters' causes but also on their mode of action.

By all these, there was practically proved the reliability of the algebraic-spectral approach to explain the eventually "statistically paradoxes" as well to overcome them in a systematically analytical manner. Still, a more general fundament for the algebraic QSAR approach containing statistical descriptors, validity, and predictability effects has to be studied at the pure mathematical level regardless of the chemical or biological systems involved.

CONCLUSION

Since the QSAR statistical analysis may often provide confusion in assessing the most reliable validated-predicted model the alternative algebraic-spectral approach was here undertaken. It is based on simple although meaningful mathematical concept of norm, which is on its turn based on the scalar product concept in a N-generalized vectorial space associate with the number of chemical compounds in a structure-activity correlation. The norm concept permits the introduction of the algebraic correlation factor seen as the ratio of the predicted to the observed norm computed from the activity values measured for the molecular toxicants considered. Moreover, it leads with formulation of the practical least action principle with the selective role among all connecting paths across uni-, bi-, and multi-parameters models computed. Such description is completed by the so called SPECTRAL-SAR determinant, built on the transformation between the direct and orthogonal vectorial space of activity-descriptors space [12], to produce a unitary QSAR analysis providing the regression equation, as well as the validation and selection framework for identifying the optimum predictive models. Nevertheless, the SPECTRAL-SAR description is based only on norm, algebraic correlation index, and on computed paths with the least ordering principle; it may eventually replace the statistical arsenal of indices as correlation factor, standard error of estimation, explained variance, Student-t, and Fisher indices and of associate tests. The success of such approach was here illustrated at the level of ecotoxicity of aliphatic amines on rats while more general studies are necessary in order to clearly asses the algebraic-spectral superiority and efficiency on the traditional or derivate statistical tools and methods. While in current mathematical formalisms the algebra is called to systematize, classify, and order the various concepts and objects for relating and unifying them in a unitary system of comprehension, we believe that algebraic-spectral approaches stand as the future in QSAR analysis trying to simplify and clarify the role, the impact and the way of action of various structural indices and parameters used in modeling observable effects. Such endeavor is currently in progress and will be reported in years to come.

KEYWORDS

- **Cyclohexylamine**
- **Endoplasmatic reticulum**
- **Ergodic**
- **Multicolinearity**

PERMISSIONS

This Chapter was previously published in Journal of Theoretical and Computational Chemistry, 8(6) (2009) as Spectral vs. Static Approach of Structure-Activity Relationship. Application on Ecotoxicity of Aliphatic Amines (by Mihai V. Putz et al.)

References

1

1. Pogliani, L. Numbers zero, one, two, and three in science and humanities. *Mathematical chemistry monographs*. University of Kragujevac-Faculty of Science, Kragujevac, 2006, Vol. 2.

2. Putz, M.V. Systematic formulation for electronegativity and hardness and their atomic scales within density functional softness theory. *Int. J. Quantum Chem.* **2006**, *106*, 361–386.

3. Putz, M.V. Semiclassical electronegativity and chemical hardness. *J. Theor. Comp. Chem.* **2007**, *6(1.1)*, 33–47.

4. Delaney, J.S.; Mullaley, A.; Mullier, G.W.; Sexton, G.J.; Taylor, R.; and Viner, R.C. Rapid construction of data tables for quantitative structure-activity relationship studies. *J. Chem. Inf. Comput. Sci.* **1993**, *33*, 174–178.

5. Klopman, G.; Balthasar, D.M.; and Rosenkranz, H.S. Application of the computer-automated structure evaluation (CASE) program to the study of structure-biodegradation relationships of miscellaneous chemicals. *Environ. Toxicol. Chem.* **1993**, *12*, 231–240.

6. Basketter, D.; Dooms-Goossens, A.; Karlberg, A.-T.; and Lepoittevin, J.P. The chemistry of contact allergy: Why is a molecule allergenic? *Contact Dermatitis* **1995**, *32*, 65–73.

7. Feijtel, T.C.J. Evaluation of the use of QSARs for priority settings and risk assessment. *SAR and QSAR Environ. Res.* **1995**, *3*, 237–245.

8. Hermens, J.L.M. and Verhaar, H.J.M. QSARs in environmental toxicology and chemistry. *ACS Symp. Ser.* **1995**, *606*, 130–140.

9. Hermes, J. Prediction of environmental toxicity based on structure-activity relationships using mechanistic information. *Sci. Total Environ.* **1995**, *171*, 235–242.

10. Hermens, J.; Balaz, S.; Damborsky, J.; Karcher, W.; Müller, M.; Peijnenburg, W.; Sabljic, A.; and Sjöström, M. Assessment of QSARs for predicting fate and effects of chemicals in the environment: An international European project. *SAR and QSAR Environ. Res.* **1995**, *3*, 223–236.

11. Ogihara, N. Drawing out drugs. *Mod. Drug Discov.* **2003**, *6(1.9)*, 28–32.

12. Hansch, C.; Hoekman, D.; and Gao, H. Comparative QSAR: Toward a deeper understanding of chemicobiological interactions. *Chem. Rev.* **1996**, *96*, 1045–1075.

13. Kubinyi, H. Der Schlüssel zum Schloß I. Grundlagen der Arzneimittelwirkung. *Pharmazie in unserer Zeit* **1994**, *23 Jahrg. Nr.3*, 158–168.

14. Liwo, A.; Tarnowska, M.; Grzonka, Z.; and Tempczyk, A. Modified free-Wilson method for the analysis of biological activity data. *Computers Chem.* **1992**, *16*, 1–9.

15. Schmidli, H. Multivariate prediction for QSAR. *Chemom. Intell. Lab. Sys.* **1997**, *37*, 125–134.

16. Lhuguenot, J.C. Relation quantitative structure-activité (QSAR): Une méthode mal reconnue car trop souvent mal utilisée. *Ann. Fals. Exp. Chim.* **1995**, *88*, 293–310.

17. Crippen, G.M.; Bradley, M.P.; and Richardson, W.W. Why are binding-site models more complicated than molecules? *Perspect. Drug Dis. Des.* **1993**, *1*, 321–328.

18. Kier, L.B. and Hall, L.H. *Molecular connectivity in structure-activity analysis*. Research Studies Press, Letchworth, 1986.

19. Balaban, A.T.; Motoc, I.; Bonchev, D.; and Mekenyan, O. Topological indices for structure-activity correlations. *Top. Curr. Chem.* **1983**, *114*, 21–55.

20. Navia, M.A. and Peattie, D.A. Structure-based drug design: Applications in immunopharmacology and immunosuppression. *Immunology Today* **1993**, *14*, 296–302.

21. Perkins, T.D.J. and Dean, P.M. An exploration of a novel strategy for superposing several flexible molecules. *J. Comput.-Aided Mol. Des.* **1993**, *7*, 155–172.

22. Lemmen, C. and Lengauer, T. Time-efficient flexible superposition of medium-sized molecules. *J. Comput.-Aided Mol. Des.* **1997**, *11*, 357–368.

23. Balaban, A.T.; Chiriac, A.; Motoc, I.; and Simon, Z. *Steric fit in QSAR*. Springer, Berlin (Lecture Notes in Chemistry Series), 1980.

24. Simon, Z.; Chiriac, A.; Holban, S.; Ciubotariu, D.; and Mihalas, G.I. *Minimum steric difference: The MTD method for QSAR studies*. Research Studies Press (Wiley), Letchworth, 1984.

25. Duda-Seiman C.; Duda-Seiman D.; Dragoş D.; Medeleanu M.; Careja V.; Putz M.V.; Lacrămă A.M.; Chiriac A.; Nuţiu R.; and Ciubotariu D. Design of anti-HIV ligands by means of minimal topological difference (MTD) method. *Int. J. Mol. Sci.* **2006**, *7*, 537–555.

26. Cramer, R.D.III; Patterson, D.E.; and Bunce, J.D. Comparative molecular field analysis (CoMFA). 1. Effect shape on binding of steroids to carrier proteins. *J. Am. Chem. Soc.* **1988**, *110*, 5959–5967.

27. Cramer, R.D.III; DePriest, S.A.; Patterson, D.E.; and Hecht, P. The developing practice of comparative molecular field analysis. In *3D QSAR in drug design. Theory, methods and applications*, H. Kubinyi (Ed.). Escom, Leiden, 1993, pp. 443–485.

28. Sun, J.; Chen, H.F.; Xia, H.R.; Yao, J.H.; and Fan, B.T. Comparative study of factor Xa inhibitors using molecular docking/SVM/HQSAR/3D-QSAR methods. *QSAR Comb. Sci.* **2006**, *25*, 25–45.

29. Randić, M.; Jerman-Blazić, B.; and Trinajstić, N. Development of 3-dimensional molecular descriptors. *Comput. Chem.* **1990**, *14*, 237–246.

30. Randić, M. and Razinger, M. Molecular topographic indices. *J. Chem. Inf. Comput. Sci.* **1995**, *35*, 140–147.

31. Manallack, D.T. and Livingstone, D.J. Artificial neural networks: Application and chance effects for QSAR data analysis. *Med. Chem. Res.* **1992**, *2*, 181–190.

32. Manallack, D.T. and Livingstone, D.J. Limitations of functional-link nets as applied to QSAR data analysis. *Quant. Struct-Act. Relat.* **1994**, *13*, 18–21.

33. Marchant, C.A. and Combes, R.D. Artificial intelligence: The use of computer methods in the prediction of metabolism and toxicity. In *Bioactive compound design: Possibilities for industrial use*, M.G. Ford and R. Greenwood (Eds.). G.T. Brooks and R. Franke BIOS Scientific Publishers Limited, Oxford, 1996.

34. Moriguchi, I.; Hirono, S.; Matsushita, Y.; Liu, Q.; and Nakagome, I. Fuzzy adaptive least squares applied to structure-activity and structure-toxicity correlations. *Chem. Pharm. Bull.* **1992**, *40*, 930–934.

35. Moriguchi, I. and Hirono, S. Fuzzy adaptive least squares and its use in quantitative structure-activity relationships. In *QSAR and drug design—New developments and applications*, T. Fujita (Ed.). Elsevier Science B. V., 1995.

36. Vapnik, V.N. *Statistical learning theory.* John Wiley & Sons, New York, 1998.

37. Vapnik, V.N. *Estimation of dependencies based on empirical data.* Springer-Verlag, Berlin, 1982.

38. Schölkpof, B.; Burges, C.J.C.; and Smola, A.J. (Eds.) *Advances in kernel methods. Support vector learning.* MIT Press, Cambridge, MA, 1999.

39. Schölkpof, B. and Smola, A.J. *Learning with kernels.* MIT Press, Cambridge, MA, 2002.

40. Mangasarian, O.L. and Musicant, D.R. Succesive overrelaxation for support vector machines. *IEEE Trans. Neural Networks* **1999**, *10*, 1032–1036.

41. Mattera, D.; Palmieri, F.; and Haykin, S. Simple and robust methods for support vector expansions. *IEEE Trans. Neural Networks* **1999**, *10*, 1038–1047.

42. Luan, F.; Ma, W.P.; Zhang, X.Y.; Zhang, H.X.; Liu, M.C.; Hu, Z.D.; and Fan, B.T. QSAR study of polychlorinated dibenzodioxins, dibenzofurans, and biphenyls using the heuristic method and support vector machine. *QSAR Comb. Sci.* **2006**, *25*, 46–55.

43. Sutter, J.M.; Kalivas, J.H.; and Lang, P.K. Which principal components to utilize for principal component regression. *J. Chemometr.* **1992**, *6*, 217–225.

44. Nendza, M. and Wenzel, A. Statistical approach to chemicals classification. *Sci. Total Environ.* **1993**, *Supplement*, 1459–1470.

45. Cash, G.G. and Breen, J.J. Principal component analysis and spatial correlation: Environmental analytical software tools. *Chemosphere* **1992**, *24*, 1607–1623.

46. Hemmateenejad, B.; Miri, R.; Jafarpour, M.; Tabarzad, M.; and Foroumadi, A. Multiple linear regression and principal component analysis-based prediction of the anti-tuberculosis activity of some 2-aryl-1,3,4-thiadiazole derivatives. *QSAR Comb. Sci.* **2006**, *25*, 56–66.

47. Randić, M. Resolution of ambiguities in structure-property studies by use of orthogonal descriptors. *J. Chem. Inf. Comput. Sci.* **1991**, *31*, 311–320.

48. Randić, M. Orthogonal molecular descriptors. *New J. Chem.* **1991**, *15*, 517–525.

49. Amić, D.; Davidović-Amić, D.; and Trinajstić, N. Calculation of retention times of anthocyanins with orthogonalized topological indices. *J. Chem. Inf. Comput. Sci.* **1995**, *35*, 136–139.

50. Lučić, B.; Nikolić, S.; Trinajstić, N.; and Juretić, D. The structure-property models can be improved using the orthogonalized descriptors. *J. Chem. Inf. Comput. Sci.* **1995**, *35*, 532–538.

51. Lučić, B.; Nikolić, S.; Trinajstić, N.; Jurić, A.; and Mihalić, Z. A structure-property study of the solubility of aliphatic alcohols in water. *Croatica Chem. Acta* **1995**, *68*, 417–434.

52. Lučić, B.; Nikolić, S.; Trinajstić, N.; Juretić, D.; and Jurić, A.A. Novel QSPR approach to physicochemical properties of the α-amino acids. *Croatica Chem. Acta* **1995**, *68*, 435–450.

53. Šoškić, M.; Plavšić, D.; and Trinajstić, N. Link between orthogonal and standard multiple linear regression models. *J. Chem. Inf. Comput. Sci.* **1996**, *36*, 829–832.

54. Klein, D.J.; Randić, M.; Babić, D.; Lučić, B.; Nikolić, S.; and Trinajstić, N. Hierarchical orthogonalization of descriptors. *Int. J. Quantum Chem.* **1997**, *63*, 215–222.

55. Ivanciuc, O.; Taraviras, S.L.; and Cabrol-Bass, D. Quasi-orthogonal basis sets of molecular graph descriptors as chemical diversity measure. *J. Chem. Inf. Comput. Sci.* **2000**, *40*, 126–134.

56. Putz M.V. A Spectral approach of the molecular structure—Biological activity relationship Part I. The general algorithm. *Ann. West Univ. Timişoara Ser. Chem.* **2006**, *15*, 159–166.

57. Putz M.V. and Lacrămă A.M. A spectral approach of the molecular structure—Biological activity relationship Part II. The enzymatic activity. *Ann. West Univ. Timişoara Ser. Chem.* **2006**, *15*, 167–176.

58. Danko P.E.; Popov A.G.; and Kozhevnikova T.Y.A. *Higher mathematics in problems and exercises.* Mir Publishers, Moscow, Vol. I, 1983.

59. Carnahan B.; Luther H.A.; and Wilkes J. *Applied numerical methods.* John Wiley & Sons, New York, 1969.

60. Young D. and Gregory R. *A survey of numerical mathematics.* Addison-Wesley, Massachusets, 1973, Vol. I, II.

61. Fadeeva V.N. *Computational methods of linear algebra.* Dover Publications, New York, 1959.

62. Daudel R.; Leroy G.; Peeters D.; and Sana M. *Quantum chemistry.* John Wiley & Sons, New York, 1983.

63. European Commission. *Proposal for a Regulation of the European Parliament and of the Council Concerning the Registration, Evaluation, Authorization and Restriction of Chemicals (REACH), Establishing a European Chemicals Agency and Amending Directive 1999/45/EC and Regulations (EC) {on Persistent Organic Pollutants}.* Brussels, Belgium, 2003.

64. OECD. *Environment Directorate Joint Meetings of the Chemicals Committee and the Working Party on Chemicals, Pesticides and Biotechnology. OECD Series on Testing and Assessment. Number 49. The Report from Expert Group on (Quantitative) Structure-Activity Relationships [(Q)SAR] on the Principles for the Validation of (Q)SARs.* Paris, France, 2004.

65. U.S. Environmental Protection Agency. ECOSAR: A computer program for estimating the ecotoxicity of industrial chemicals based on structure activity relationships. *EPA 748-R-002.* National Center for

Environmental Publications and Information, Cincinnati, OH, 1994, p. 34.

66. US EPA AQUIRE (AQUatic Toxicity Information REtrival). U.S. Environmental Protection Agency, 2002. ECOTOX User Guide: ECOTOXicology Database System. Version 3.0 [http:www.epa.gov/ecotox/], 2002.

67. Greim, H.; Csanády, G.; Filser, J.G.; Kreuzer, P.; Schwarz, L.; Wolff, T.; and Werner, S. Biomarkers as tools in human health risk assessment. *Clin. Chem.* **1995**, *41*, 1804–1808.

68. Pangrekar, J.; Klopman, G.; and Rosenkranz, H.S. Expert-system comparison of structural determinants of chemical toxicity to environmental bacteria. *Environ. Toxicol. Chem.* **1994**, *13*, 979–1001.

69. Klopman, G.; Zhang, Z.; Woodgate, S.D.; and Rosenkranz, H.S. The structure-toxicity relationship challenge at hazardous waste sites. *Chemosphere* **1995**, *31*, 2511–2519.

70. Judson, P.N. QSAR and expert systems in the prediction of biological activity. *Pestic. Sci.* **1992**, *36*, 155–160.

71. Hulzebos, E. and Posthumus, R. (Q)SAR: Gatekeepers against risk on chemicals? *SAR and QSAR Environ. Res.* **2003**, *14*, 285–316.

72. Hulzebos, E.; Sijm, D.; Traas, T.; Posthumus, R.; and Maslankiewicz, L. Validity and validation of expert (Q)SAR systems. *SAR and QSAR Environ. Res.* **2005**, *16*, 385–401.

73. Pavan, M.; Netzeva, T.I.; and Worth, A.P. Validation of a QSAR model for acute toxicity. *SAR and QSAR Environ. Res.* **2006**, *17*, 147–171.

74. Cronin, M.T.D. and Dearden, J.C. QSAR in toxicology. 1. Prediction of aquatic toxicity. *Quant. Struct.-Act. Relat.* **1995**, *14*, 1–7.

75. Cronin, M.T.D. and Dearden, J.C. QSAR in toxicology. 2. Prediction of acute mammalian toxicity and interspecies correlations. *Quant. Struct.-Act. Relat.* **1995**, *14*, 117–120.

76. Cronin, M.T.D. and Dearden, J.C. QSAR in toxicology. 3. Prediction of chronic toxicities. *Quant. Struct.-Act. Relat.* **1995**, *14*, 329–334.

77. Cronin, M.T.D. and Dearden, J.C. QSAR in toxicology. 4. Prediction of non-lethal mammalian toxicological endpoints, and expert systems for toxicity prediction. *Quant. Struct.-Act. Relat.* **1995**, *14*, 518–523.

78. Topliss, J.G. and Costello, J.D. Chance correlation in structure-activity studies using multiple regression analysis. *J. Med. Chem.* **1972**, *15*, 1066–1069.

79. Cronin, T.D.; Aptula, A.O.; Duffy, J.C.; Netzeva, T.I.; Rowe, P.H.; Valkova, I.V.; and Wayne-Schultz, T. Comparative assessment of methods to develop QSARs for the prediction of the toxicity of phenols to *Tetrahymena pyriformis*. *Chemosphere* **2002**, *49*, 1201–1221.

80. Lynn, D.H. and Small, E.B. Phylum ciliophora. In *Handbook of protoctista*, L. Margulis, J.O. Corliss, M. Melkonian, and D.J. Chapman (Eds.). Jones and Bartlett Publishers, Boston, 1991.

81. About systematic classification of *Tetrahymena pyriformis*:
    (a) http://www.ucmp.berkeley.edu/protista/ciliata/ciliatamm.html (December 2006).
    (b) http://www.ns.purchase.edu/biology/bio1560lab/protista.htm (December 2006).
    (c) http://www.bch.umontreal.ca/ogmp/projects/tpyri/org.html (December 2006).
    (d) http://www-micro.msb.le.ac.uk/video/Tetrahymena.html (December 2006).

82. Niles, E.G. and Jain, R.K. Physical map of the ribosomal ribonucleic acid gene from *Tetrahymena pyriformis*. *Biochemistry* **1981**, *20*, 905–909.

83. Manasherob, R.; Ben-Dov, E.; Zaritsky, A.; and Barak, Z. Germination, growth, and sporulation of *Bacillus thuringiensis* subsp. *israelensis* in excreted food vacuoles of the protozoan *Tetrahymena pyriformis*. *Appl. Environ. Microbiol.* **1998**, *64*, 1750–1758.

84. Strüder-Kypke, M.C.; Wright, A.-D.G.; Jerome, C.A.; and Lynn, D.H. Parallel evolution of histophagy in ciliates of the genus *Tetrahymena*. *BMC Evol. Bio.* **2001**, *1*:5.

85. Putz M.V.; Russo N.; and Sicilia E. Atomic radii scale and related size properties from density functional electronegativity formulation. *J. Phy. Chem. A* **2003**, *107*, 5461–5465.

86. Cronin, M.T.D.; Netzeva, T.I.; Dearden, J.C.; Edwards, R.; and Worgan A.D.P. Assessment and modeling of the toxicity of organic chemicals to *Chlorella vulgaris*: Development of a novel database. *Chem. Res. Toxicol.* **2004**, *17*, 545–554.

87. Schultz T.W. TETRATOX: The *Tetrahymena pyriformis* population growth impairment endpoints. A surrogate for fish lethality. *Toxicol. Methods* **1997**, *7*, 289–309.

88. Schultz T.W. Structure-toxicity relationships for benzene evaluated with *Tetrahymena pyriformis*. *Chem. Res. Toxicol.* **1999**, *12*, 1262–1267.

89. Schultz T.W.; Cronin M.T.D.; Netzeva T.I.; and Aptula A.O. Structure-toxicity relationships for aliphatic chemicals evaluated with *Tetrahymena pyriformis*. *Chem. Res. Toxicol.* **2002**, *15*, 1602–1609.

90. Schultz T.W.; Netzeva T.I.; and Cronin M.T.D. Selection of data sets for QSARs: Analyses of *Tetrahymena* toxicity from aromatic compounds. *SAR and QSAR Environ. Res.* **2003**, *14*, 59–81.

91. Hypercube, Inc. (2002) HyperChem 7.01 [Program package].

92. StatSoft, Inc. (1995). STATISTICA for Windows [Computer program manual].

2

1. Simon, Z.; Chiriac, A.; and Ostafe, V. *Accuracy in molecular biologic recognition*. Mirton Publishing House, Timişoara, 2000.

2. Henri, V. Über das gesetz der wirkung des invertins. *Z. Phys. Chem.* **1901**, *39*, 194–216.

3. Michaelis, L. and Menten, M.L. Die kinetik der invertinwirkung, *Biochem. Z.* **1913**, *49*, 333–369.

4. Chiriac, A.; Ciubotariu, D.; and Simon, Z. (Eds.). *Relaţii cantitative structură chimică activitate biologică (QSAR), Metoda MTD*. Mirton Publishing House, Timişoara, 1996, Chapters 1, 5.

5. Oprea, T.I.; Kurunczi, L.; and Martin, O. *Curs de proiectarea moleculară a medicamentului*. Mirton Publishing House, Timişoara, 1999.

6. Putz, M.V.; Lacrămă, A.M.; and Ostafe, V. Full analytic progress curves of the enzymic reactions in vitro. *Int. J. Mol. Sci.* **2006**, *7*, 469–484.

7. Chiriac, A.; Mracec, M.; Oprea, T.I.; Kurunczi, L.; and Simon, Z. (Eds.). *Quantum biochemistry and specific interaction*. Mirton Publishing House, Timişoara, 2003, Chapter 8.

8. Felszeghy, E. and Abraham, A. *Biochemistry* (in Romanian). Didactic and Pedagogic Publishing House, Bucharest, 1972, Chapter 5.

9. Kurunczi, L. *Computer assisted drug design. QSAR* (in Romanian). University of Medicine and Pharmacy of Timişoara, ISBN: 973-9336-66-3, 1998, Chapter 4.

10. Miller, J.N. and Miller J.C. *Statistics and chemometrics for analytical chemistry*, 4th edn. Pretience Hall, Harlow, 2000.

11. Putz, M.V. A spectral approach of the molecular structure-biological activity relationship Part I. The general algorithm. *Ann. West Univ. Timişoara Ser. Chem.* **2006**, *15,*159–166.

12. Silver, A. *The Biology of Cholinesterases*. North Holland Publishing Co. Ltd., Amsterdam, 1974.

13. Massoulié, J.; Pezzementi, L.; Bon, S.; Kreja, E.; and Valette, F.M. Molecular and cellular biology of cholinesterases. *Prog. Neurobiol.* **1993**, *41*, 31–91.

14. Järv, J.; Kesvatera, T.; and Aaviksaar; A. Structure-activity relationships in acetylcholinesterase reactions. Hydrolysis of non-ionic acetic esters. *Eur. J. Biochem.* **1976**, *67*, 315–322.

15. Blaga, P. *Statistics ... through MATLAB* (in Romanian). Cluj University Press-Cluj-Napoca (Romania), 2002, Annex. IV.

16. Gray, P. and Scott, S.K. *Chemical oscillations and instabilities. Non-linear Chemical Kinetics*. Clarendon Press, Oxford, 1990.

17. Segel, I.H. *Enzyme kinetics: Behavior and analysis of rapid equilibrium and steady-state systems*. Wiley, New York, 1975.

18. Segel, L.A. On the validity of the steady state assumption of enzyme kinetics. *Bull. Math. Biol.* **1988**, *50*, 579–593.

19. Segel, L.A. and Slemrod, M. The quasi-steady-state assumption: A case study in perturbation. *SIAM Rev.* **1989**, *31*, 446–477.

3

1. Cronin, M.T.D. and Dearden, J.C. QSAR in toxicology. 1. Prediction of aquatic toxicity. *Quant. Struct.-Act. Relat.* **1995**, *14*, 1–7.

2. Schultz, T.W. and Mekenyan, O.G. *Responce-surface analyses: A comparison of two approaches to predicting acute toxicity in quantitative structure activity relationship in environmental sciences*, VIII. SE-TAC Press, Pensacola, FL, 2002.

3. Repetto, G.; Jos, A.; Hazen, M.J.; Molero, M.L.; del Peso, A.; Salguero, M.; del Castillo, P.; Rodriguez-Vicente, M.C.; and Repetto, M. A test battery for the ecotoxicological evaluation of pentachlorophenol. *Toxicology in Vitro* **2001**, *15(4–5)*, 503–509.

4. Johnson, I.; Hutchings, M.; Benstead, R.; Thain, J.; and Whitehouse, P. Bioassay selection, experimental design and quality control/assurance for use in effluent assessment and control. *Ecotoxicology* **2004**, *13*, 437–447.

5. Ahlers, J. Strategies for risk assessment of existing chemicals in soil. *J. Soils & Sediments* **2001**, *1*, 168–174.

6. Ahlers, J. and Martin, S. Global Soils: EU, Risk assessment of chemicals in soil: recent developments in the EU. *J. Soils & Sediments* **2003**, *3(4)*, 240–241.

7. Hund-Rinke, K.; Kordel, W.; Hennecke, D.; Eisentrager, A.; and Heiden, St. Bioassays for the ecotoxicological and genotoxicological assessment of contaminated soils (Results of a Round Robin test). Part I. Assessment of a possible groundwater contamination: ecotoxicological and genotoxicological tests with aqueous soil extracts. *J. Soils & Sediments* **2002**, *2*, 43–50.

8. Hund-Rinke, K.; Kordel, W.; Hennecke, D.; Achazi, R.K.; Warnecke, D.; Wilke, B.M.; and Heiden, St. Bioassays for the ecotoxicological and genotoxicological assessment of contaminated soils (Results of a Round Robin test). Part II. Assessment of the habitat function of soils: tests with soil microflora and fauna. *J. Soils & Sediments* **2002**, *2*, 83–90.

9. Achazi, R.K. Invertebrates in risk assessment development of a test battery and of short term biotests for ecological risk assessment of soil. *J. Soils & Sediments* **2002**, *2(4)*, 174–178.

10. Cronin, M.T.D.; Netzeva, T.I.; Dearden, J.C.; Edwards, R.; and Worgan, A.D.P. Assessment and modeling of the toxicity of organic chemicals to *Chlorella vulgaris*: Development of a novel database. *Chem. Res. Toxicol.* **2004**, *17*, 545–554.

11. Jastorff, B.; Stormann, R.; Ranke, J.; Molter, K.; Stock, F.; Oberheitmann, B.; Hoffmann, W.; Nuchterm, M.; Ondruschka, B.; and Filser, J. How hazardous are ionic liquids? Structure-activity relationship and biologic testing as important elements for sustainability evaluation. *Green Chemistry* **2003**, *5*, 136–142.

12. Mincea, M.; Lacrămă, A.M.; Stoian, C.; Baicu, I.; Nemes, N.; Popet, L.; and Ostafe, V. Multienzimatic test battery—A model for testing at molecular level the toxicity of chemical compounds. *Sustainability for Humanity and Environment in the Extended Connection Field Science-Economy.* Polytechnic Publishing House, Timişoara, 2005, Vol. II, pp. 211–214.

13. Putz, M.V.; Lacrămă, A.M.; and Ostafe, V. Introducing logistic enzyme kinetics. *J. Optoel. Adv. Mat.* **2007**, 2910–2916.

14. Putz, M.V.; Lacrămă, A.M.; and Ostafe, V. Full analytic progress curves of enzymic reaction in vitro. *Int. J. Mol. Sci.* **2006**, *7*, 417–433.

15. Putz, M.V.; Lacrămă, A.M.; and Ostafe, V. Full time course analysis for reversible enzyme kinetics. In *Proceedings of the VIIIth International Symposium YOUNG PEOPLE AND MULTIDISCIPLINARY RESEARCH Romania–Serbia–Hungary.* Welding Publishing House, Timişoara, 2006, pp. 642–649.

16. Cronin, M.T.D.; Dearden, J.C.; Walker, J.D.; and Worth, A.P. Quantitative structure-activity relationships for human health effects: Commonalities with other endpoints. *Environ. Toxicol. Chem.* **2003**, *22(8)*, 1829–1843.

17. Cronin, M.T.D. Computational methods for the prediction of drug toxicity. *Curr. Opin. Drug. Discov. Dev.*, **2000**, *3*, 292–297.

18. Schultz, T.W.; Cronin, M.T.D.; Walker, J.D.; Aptula, A.O. Quantitative structure-activity relationships (QSARs) in toxicology: A historical perspective. *J. Mol. Struct. (Theochem.)*, **2003**, *622*, 1–22.

19. Schultz, T.W.; Netzeva, T.I.; and Cronin, M.T.D. Selection of data sets for QSARs: Analyses of *Tetrahymena* toxicity from aromatic compounds. *SAR and QSAR Environ. Res.*, **2003**, *14(1)*, 59–81.

20. Saliner, A.G.; Amat, L.; Carbo-Dorca, R.; Schultz, T.W.; and Cronin, M.T.D. Molecular quantum similarity analysis of estrogenic activity. *J. Chem. Inf. Comput. Sci.*, **2003**, *43*, 1166–1176.

21. Cronin, M.T.D. and Schultz, T.W. Pitfals in QSAR. *J. Mol. Struct. (Theochem.)*, **2003**, *622*, 39–51.

22. Netzeva, T.I.; Dearden, J.C.; Edwards, R.; Worgan, A.D.P.; and Cronin, M.T.D. QSAR analysis of the toxicity of aromatic compounds to *Chlorella vulgaris* in a novel short-term assay. *J. Chem. Inf. Comput. Sci.* **2004**, *44*, 258–265.

23. Schultz, T.W. Structure-toxicity relationships for benzene evaluated with *Tetrahymena pyriformis*. *Chem. Res. Toxicol.* **1999**, *12*, 1262–1267.

24. Dearden, J.C. In *Practical applications of quantitative structure-activity relationships (QSAR) in environmental chemistry and toxicology*. EEC, Brussels, 1990, pp. 25–59.

25. Livingstone, D.J. The characterization of chemical structures using molecular properties. A survey. *J. Chem. Inf. Comput. Sci.* **2000**, *40*, 195–209.

26. Benigni, R.; Cotta-Ramusino, M.; Gallo, G.; Giorgi, F.; Giuliani, A.; and Vari, M.R. Deriving a quantitative chirality measure from molecular similarity indices. *J. Med. Chem.* **2000**, *43*, 3699–3703.

27. Cronin, M.T.D. *Molecular descriptors of QSAR*. Proceedings of the seminar of current topics in toxicology: QSAR in Toxicology. T. Coccini, L. Giannoni, W. Karcher, L. Manzo, R. Roi (Eds.). Commission of the European Communities, Luxembourg, 1992, pp. 43–54.

28. Tute, M.S. History and objectives of quantitative drug design. In *Comprehensive medicinal chemistry. Quantitative drug design*, C.A. Ramsden (Ed.). Pergammon Press, Oxford, 1990, Vol. 4, pp. 1–3.

29. Patel, H. and Cronin, M.T.D. A novel index for the description of molecular linearity. *J. Chem. Inf. Comput. Sci.* **2001**, *41*, 1228–1236.

30. Cronin, M.T.D. and Dearden, J.C. QSAR in toxicology. 3. Prediction of chronic toxicities. *Quant. Struct. Act. Relat.* **1995**, *14*, 329–334.

31. Schultz, T.W.; Cronin, M.T.D.; Netzeva, T.I.; and Aptula, A.O. Structure-toxicity relationships for aliphatic chemicals evaluated with *Tetrahymena pyriformis*. *Chem. Res. Toxicol.* **2002**, *15*, 1602–1609.

32. Ramos, E.U.; Vaes, W.H.J.; Mayer, P.; and Hermens, J.L.M. Algal growth inhibition of *Chlorella pyrenoidosa* by polar narcotic pollutants: Toxic cell concentrations and QSAR modelling. *Aquatic Toxicol.* **1999**, *46*, 1–10.

33. Moss, G.P.; Dearden, J.C.; Patel, H.; and Cronin, M.T.D. Quantitative structure-permeability relationships (QSPRs) for precutaneous absorbtion. *Toxicology in Vitro*, **2002**, *16*, 299–317.

34. Patel, H.; Berge, W.; and Cronin, M.T.D. Quantitative structure-activity relationships (QSARs) for the prediction of skin permeation of exogenous chemicals. *Chemosphere*, **2002**, *48*, 603–613.

35. Patel, H.; Schultz, T.W.; and Cronin, M.T.D. Physico-chemical interpretation and prediction of the dimyristoyl phosphatidyl choline-water partition coefficient. *J. Mol. Struct. (Theochem.)* **2002**, *593*, 9–18.

36. Schultz, T.W. and Cronin, M.T.D. Essential and desirable characteristics of ecotoxicity quantitative structure-toxicity relationships. *Environ. Toxicol. Chem.* **2003**, *22*, 599–607.

37. Hulzebos, E.M. and Posthumus, R. (Q)SARs: Gatekeepers against risk on chemicals? *SAR and QSAR Environ. Res.* **2003**, *14(4)*, 285–316.

38. Bradbury, S.P.; Russom, C.L.; Ankley, G.T.; Schultz, T.W.; and Walker, J.D. Overview of data and conceptual approaches for derivation of quantitative structure-activity

relationships for ecotoxicological effects of organic chemicals. *Environ. Toxicol. Chem.* **2003**, *22(8)*, 1789–1798.

39. Hermens, J.L.M. and Verhaar, H.J.M. QSAR in environmental toxicology and chemistry. *Classical and three-dimensional QSAR in agrochemistry*, C. Hansch, and T. Fujita (Eds.). American Chemical Society, Washington, DC, 1995, Vol. 10.

40. Bradbury, S.P. Quantitative structure-activity relationships and ecological risk assessment: An overview of predictive aquatic toxicology research. *Toxicol. Lett.* **1995**, *79*, 229–237.

41. About systematic classification of *Chlorella vulgaris*: http://en.wikipedia.org/wiki/Chlorella (Accessed 06.12.2006).

42. Patrut, D.I. *Botanica Sistematica, Thalobionta et Bryobionta* (in Romanian). Aprilia Print Publisher, Timişoara, 2004, 17–37.

43. Dittrich, M. and Obst, M. Are picoplankton responsible for calcite precipitation in lakes? *Ambio* **2004**, *33*, 559–564.

44. Happey-Wood, C.M. Ecology of freshwater planktonic green algae, *Growth and Reproductive Strategies of Freshwater Phytoplankton*, C.D. Sandgreen (Ed.). Cambridge University Press, Cambridge, 1988, Vol. 5, p. 175.

45. Becker, E.W. *Microalgae: Biotechnology and Microbiotechnology*. Cambridge University Press, Cambridge, 1994.

46. Worgan, A.D.P.; Dearden, J.C.; Edwards, R.; Netzeva, T.I.; and Cronin, M.T.D. Evaluation of a novel short-term algal toxicity assay by the development of QSARs and inter-species relationships for narcotic chemicals. *QSAR Comb. Sci.* **2003**, *22*, 204–209.

47. Leszczynska, M. and Oleszkiewik, J.A. Application of fluorescein diacetate hydrolysis as an acute toxicity test. *Environ. Technol.* **1996**, *17*, 79–85.

48. About systematic classification of *Vibrio fischeri*:
    (a) http://en.wikipedia.org/wiki/Bobtail_squid (Accessed December 2006).
    (b) http://en.wikipedia.org/wiki/Vibrio_fischeri (Accessed December 2006).
    (c) http://www.infoplease.com/ce6/sci/A0863233.html (Accessed December 2006).

    (d) http://lanesville.k12.in.us/LCSYellowpages/Tickit/Carl/bacteria.html (Accessed December 2006).

49. Millikan, D.S. and Ruby, E.G. Alterations in *Vibrio fischeri* motility correlate with a delay in symbiosis initiation and are associated with additional symbiotic colonization defects. *Appl. Env. Microbiol.* **2002**, *68*, 2519–2528.

50. Holt, J.G. (Ed.) *Bergey's manual of determinative bacteriology*, 9th ed. Williams and Wilkins, Baltimore, 1994.

51. Madigan, M. and Martinko, J. *Brock Biology of Microorganisms*, 11th ed. Prentice Hall, 2005.

52. Ruby, G.; Urbanowski, M.; Campbell, J.; Dunn, A.; Faini, M.; Gunsalus, M.; Lostroh, M.; Lupp, C.; McCann, J.; Millikan, D.; Schaefer, A.; Stabb, E.; Stevens, A.; Visick, K.; Whistler, C.; and Greenberg, E. Complete genome sequence of vibrio fischeri: A symbiotic bacterium with pathogenic congeners. *PNAS* **2005**, *102(8)*, 3004–3009.

53. Kaiser, K.L.E. and Palabrica, V.S. *Photobacterium phosphoreum* toxicity data index. *Water Pollut. Res. J. Can.* **1991**, *26*, 361–431.

54. ISO 11348-3. Standard method Microtox, *International Standard Organisation.*1998.

55. About the systematic classification of *Pimephales promelas*:

    http://www.gen.umn.edu/research/fish/fishes/fathead_minnow.html (Accessed December 2006).

    http://www.rook.org/earl/bwca/nature/fish/pimephalespro.html (Accessed December 2006).

    http://mblaquaculture.com (Accessed December 2006).

56. Russom, C.L.; Bradbury, S.P.; Broderius, S.J.; Hammermeister, D.E.; and Drummond, R.A. Predicting models of toxic action from chemical structure: Acute toxicity in the fathead minnow (*Pimephales promelas*). *Environ. Toxicol. Chem.* **1997**, *16*, 948–967.

57. (a) Putz, M.V. A Spectral approach of the molecular structure—Biological activity relationship Part I. The general algorithm, *Ann. West Univ. Timişoara Ser. Chem.* **2006**, *15*, 159–166.

(b) Putz, M.V.; Lacrămă, A.M. A spectral approach of the molecular structure—Biological activity relationship Part II. The enzymatic activity, *Ann. West Univ. Timişoara Ser. Chem.* **2006**, *15*, 167–176.

58. Putz, M.V.; Lacrămă, A.M. Introduction spectral structure-activity relationship (S-SAR) analysis. Application to ecotoxicology. *Int. J. Mol. Sci.*, **2007**, *8*, 363–391.

59. Miller J.N. and Miller J.C. *Statistics and Chemometrics for Analytical Chemistry*, 4th ed. Prentice Hall, Harlow, 2000.

60. Hypercube, Inc. 2002, HyperChem 7.01, Program package.

61. Netzeva, T.I.; Worgan, A.D.P.; Dearden, J.C.; Edwards, R.; and Cronin, M.T.D. Toxicological evaluation and QSAR modelling of aromatic amines to *Chlorella vulgaris*. *Bull. Environ. Contam. Toxicol.* **2004**, *73*, 385–391.

62. Schultz, T.W. TETRATOX: The *Tetrahymena pyriformis* population growth impairment endpoints. A surrogate for fish lethality. *Toxicol. Methods* **1997**, *7*, 289–309.

63. European Commission. White Paper on the Strategy for a Future Chemicals Policy, Commission of the European Communities: Brussels, Belgium. *KOM* **2001**, *88*.

64. Hulzebos, E.M.; Sijm, D.; Traas, T.; Posthumus, R.; and Maslankiewicz, L. Validity and Validation of Expert (Q)SAR Systems. *SAR and QSAR Environ. Res.* **2005**, *16(4)*, 385–401.

65. Pavan, M.; Netzeva, T.I.; and Worth, A.P. Validation of a QSAR model for acute toxicity. *SAR and QSAR Environ. Res.* **2006**, *17(2)*, 147–171.

66. European Commission. Proposal for a Regulation of the European Parliament and of the Council concerning the Registration, Evaluation, Authorization and Restriction of Chemicals (REACH), establishing a European Chemicals Agency and amending directive 1999/45/EC and Regulation (EC) {on Persistent Organic Pollutants}. Brussels, Belgium, 2003.

67. OECD. Annexes to the Report on Principles for Establishing the Status of Development and Validation of (Quantitative) Structure-Activity Relationships [(Q)SARs], 2004.

68. OECD, Environment Directorate Joint Meeting of the Chemicals Committee and the Working Party on Chemicals, pesticides and Biotechnology. OECD Series on Testing and Assessment. Number 49. The report from the Expert Group on (Quantitative) Structure-Activity Relationships [(Q)SARs] on the Principles for the Validation of (Q) SARs, 2004.

69. Worth, A.P. and Balls, M. Alternative (non-animal) methods for chemical testing: Current status and future aspects. *ATLA* **2002**, *30(1)*, 1–125.

70. Cronin, M.T.D.; Aptula, A.O.; Duffy, J.C.; Netzeva, T.I., Rowe, P.H.; Valkova, I.V.; and Schultz, T.W. Comparative assessment of methods to develop QSARs for the prediction of the toxicity of phenols to *Tetrehymena pyriformis*. *Chemosphere* **2002**, *49*, 1201–1221.

71. Worth, A.P. and Cronin, M.T.D. The use of discriminant analysis, logistic regression and classification tree analysis in the development of classification models for human health effects. *J. Mol. Struct. (Theochem.)* **2003**, *622*, 97–111.

72. Putz, M.V. and Putz, A.M. Logistic vs. W-Lambert information in quantum modeling of enzyme kinetics. *Int. J. Chemoinfo. Chem. Eng.* **2011**, *1*, 42–60.

73. Putz, M.V. and Putz, A.M. Logistic vs. W-Lambert information in modelling enzyme kinetics. In *Advanced methods and applications in chemoinformatics: Research methods and new applications*, E.A. Castro and A.K. Haghi (Eds.). IGI Global, Hershey, PA, 2011.

4

1. Corvini, P.F.X.; Vinken, R.; Hommes, G.; Schmidt, B.; and Dohmann M. Degradation of the radioactive and non-labelled branched 4(3′,5′-dimethyl-3′-heptyl)-phenol nonylphenol isomer by *Sphingomonas* TTNP3. *Biodegradation* **2004**, *15*, 9–18.

2. Schwitzguebel, J.-P.; Aubert, S.; Grosse, W.; and Laturnus, F. Sulphonated aromatic pollutants: Limits of microbial degradability and potential of phytoremediation. *Environ. Sci. Pollut. Res.* **2002**, *9*, 62–72.

3. Berking, S. Effects of the anticonvulsant drug valproic acid and related substances on

developmental processes in hydroids. *Toxic. In Vitro* **1991**, *5*, 109–117.

4. Chicu, S.A. and Berking S. Interference with metamorphosis induction in the marine cnidaria *Hydractinia echinata* (hydrozoa): A structure-activity relationship analysis of lower alcohols, aliphatic and aromatic hydrocarbons, thiophenes, tributyl tin and crude oil. *Chemosphere* **1997**, *34*, 1851–1866.

5. Chicu, S.A.; Herrmann, K.; and Berking S. An approach to calculate the toxicity of simple organic molecules on the basis of QSAR analysis in *Hydractinia echinata* (Hydrozoa, Cnidaria). *Quant. Struct.-Act. Relat.* **2000**, *19*, 227–236.

6. Chicu, S.A. and Simu, G.M. *Hydractinia echinata* test system. I. Toxicity determination of some benzenic, biphenilic and naphthalenic phenols. Comparative SAR-QSAR study. *Rev. Roum. Chim.* **2009**, *8*, in press.

7. Putz, M.V. and Lacrămă, A.M. Introducing spectral structure activity relationship (SPECTRAL-SAR) analysis. Application to ecotoxicology. *Int. J. Mol. Sci.* **2007**, *8*, 363–391.

8. Putz, M.V.; Lacrămă, A.M.; and Ostafe, V. SPECTRAL-SAR ecotoxicology of ionic liquids. The *Daphnia magna* case. *Res. Lett. Ecol.* **2007**, Article ID12813/5 pages, DOI: 10.1155/2007/12813.

9. Lacrămă, A.M.; Putz, M.V.; and Ostafe. V. A SPECTRAL-SAR model for the anionic-cationic interaction in ionic liquids: Application to *Vibrio fischeri* ecotoxicity. *Int. J. Mol. Sci.* **2007**, *8*, 842–863.

10. Putz, M.V.; Putz, A.M.; Lazea, M.; Ienciu, L.; and Chiriac A. Quantum-SAR extension of the SPECTRAL-SAR algorithm. Application to polyphenolic anticancer bioactivity. *Int. J. Mol. Sci.* **2009**, *10*, 1193–1214.

11. Cronin, M.T.D; Netzeva, T.I.; Dearden, J.C.; Edwards, R.; and Worgan, A.D.P. Assessment and modeling of toxicity of organic chemicals to *Chlorella vulgaris*: Development of a novel database. *Chem. Res. Toxicol.* **2004**, *17*, 545–554.

12. Cronin, M.T.D. and Dearden, J. QSAR in toxicology. 2. Prediction of acute mammalian toxicity and interspecies correlations.

Quant. Struct.-Act. Relat. **1995**, *14*, 117–120.

13. Cronin, M.T.D. and Worth, A.P. (Q)SARs for predicting effects relating to reproductive toxicity. *QSAR Comb. Sci.* **2008**, *27*, 91–100.

14. Dirac, P.A.M. *The principles of quantum mechanics.* Oxford University Press, Oxford, 1944.

15. Randić, M. Orthogonal molecular descriptors. *New J. Chem.* **1991**, *15*, 517–525.

16. Klein, D.J.; Randić, M.; Babić, D.; Lučić, B.; Nikolić, S.; and Trinajstić, N. Hierarchical orthogonalization of descriptors. *Int. J. Quantum Chem.* **1997**, *63*, 215–222.

17. Fernández, F.M.; Duchowicz, P.R.; and Castro, E.A. About orthogonal descriptors in QSPR/QSAR theories. *MATCH Commun. Math. Comput. Chem.* **2004**, *51*, 39–57.

18. Steen, L.A. Highlights in the history of spectral theory. *Amer. Math. Monthly* **1973**, *80*, 359–381.

19. Siegmund-Schultze, R. Der Beweis des Hilbert-Schmidt theorem. *Arch. Hist. Ex. Sc.* **1986**, *36*, 251–270.

20. Chaterjee, S.; Hadi, A.S.; and Price, B. *Regression analysis by examples*, 3rd ed. Wiley, New York, 2000.

21. Putz, M.V. and Putz, A.M. Timişoara spectral—Structure activity relationship (SPECTRAL-SAR) algorithm: From statistical and algebraic fundamentals to quantum consequences. In *Quantum frontiers of atoms and molecules*, M.V. Putz (Ed.). NOVA Publishers, Inc., New York, 2010.

22. Putz, M.V.; Duda-Seiman, C.; Duda-Seiman, D.M.; and Putz, A.M. Turning SPECTRAL-SAR into 3D-QSAR analysis. Application on H^+K^+-ATPase inhibitory activity. *Int. J. Chem. Mod.* **2008**, *1*, 45–62.

23. Lacrămă, A.M.; Putz, M.V.; and Ostafe, V. Designing a spectral structure-activity ecotoxico-logistical battery. In *Advances in quantum chemical bonding structures*, M.V. Putz (Ed.). Transworld Research Network, Kerala, India, 2008, Chapter 16, pp. 389–419.

24. Putz, M.V.; Putz, A.M.; Lazea, M.; and Chiriac, A. Spectral vs. statistic approach of structure-activity relationship. Application

on ecotoxicity of aliphatic amines. *J. Theor. Comput. Chem.* **2009**, *8*, 1235–1251.

25. Topliss, J.G. and Costello, R.J. Chance correlation in structure-activity studies using multiple regression analysis. *J. Med. Chem.* **1972**, *15*, 1066–1068.

26. Hypercube, Inc. *HyperChem 701, Program package, Semiempirical, AM1, Polak-Ribier optimization procedure*, 2002.

27. Hansch, C. and Zhang, L. Comparative QSAR: Radical toxicity and scavenging. Two different sides of the same coin. *SAR. QSAR. Environ. Res.* **1995**, *4*, 73–82.

28. Baratt, M.D. Prediction of toxicity from chemical structure. *Cell Biol. Toxicol.* **2000**, *16*, 1–13.

29. Bassan, A. and Worth, A.P. The integrated use of models for the properties and effects of chemicals by means of a structured workflow. *QSAR Comb. Sci.* **2008**, *27*, 6–20.

5

1. Horn, J. The proton-pump inhibitors: Similarities and differences. *Clin. Ther.* **2000**, *22*, 266–280.

2. Scarpignato, C. New drugs to suppress acid secretion: Current and future developments. *Drug Discov. Today: Ther. Strategies* **2007**, DOI: 10.1016/j.ddstr.2007.09.003.

3. Vanderhoff, B.T. and Tahboub, R.M. Proton pump inhibitors: An update. *Am. Fam. Physician.* **2002**, *66*, 273–280.

4. Sachs, G.; Shin, J.M.; and Howden, C.W. Review article: The clinical pharmacology of proton pump inhibitors. *Aliment. Pharmacol. Ther.* **2006**, *23(Suppl. 2)*, 2–8.

5. Marchetti, F.; Gerarduzzi, T.; and Ventura, A. Proton pump inhibitors in children: A review. *Digest. Liver Dis.* **2003**, *35*, 738–746.

6. Jain, K.S.; Shah, A.K.; Bariwal, J.; Shelke, S.M.; Kale, A.P.; Jagtap, J.R.; and Bhosale, A.V. Recent advances in proton pump inhibitors and management of acid-peptic disorders. *Bioorg. Med. Chem.* **2007**, *15*, 1181–1205.

7. Prabhakar, Y.S.; Solomon, V.R.; Gupta, M.K.; and Katti, S.B. QSAR studies on thiazolidines: A biologically privileged scaffold. *Top. Heterocycl. Chem.* **2006**, *4*, 161–249.

8. Putz, M.V. A spectral approach of the molecular structure—Biological activity relationship Part I. The general algorithm. *Ann. West Univ. Timişoara, Ser. Chem.* **2006**, *15*, 159–166.

9. Putz, M.V. and Lacrămă, A.M. A spectral approach of the molecular structure—Biological activity relationship Part II. The enzymatic activity, *Ann. West Univ. Timişoara, Ser. Chem.* **2006**, *15*, 167–176.

10. Putz, M.V. and Lacrămă, A.M. Introducing spectral structure activity relationship (SPECTRAL-SAR) analysis. Application to ecotoxicology. *Int. J. Mol. Sci.*, **2007**, *8*, 363–391.

11. Lacrămă, A.M.; Putz, M.V.; and Ostafe, V. A SPECTRAL-SAR model for the anionic-cationic interaction in ionic liquids: Application to *Vibrio fischeri* ecotoxicity. *Int. J. Mol. Sci.* **2007**, *8*, 842–863.

12. Putz, M.V.; Lacrămă, A.M.; and Ostafe, V. SPECTRAL-SAR ecotoxicology of ionic liquids. The *Daphnia magna* case. *Res. Lett. Ecol.* **2007**, Article ID12813/5 pages, DOI: 10.1155/2007/12813 (http://www.hindawi.com/journals/rleco/q4.2007.html)

13. Lacrămă, A.M.; Putz, M.V.; and Ostafe, V. Designing a spectral structure-activity ecotoxico-logistical battery. In *Advances in quantum chemical bonding structures*, M.V. Putz (Ed.). Transworld Network Research, Kerala, India, 2008, Chapter 16, pp. 389–419 (http://www.trnres.com/putz.htm)

14. Miller, J.N. and Miller, J.C. *Statistics and chemometrics for analytical chemistry.* 4th ed. Prentice Hall, New York, 2000.

15. Putz, M.V. and Putz, A.M. Timişoara spectral—Structure activity relationship (SPECTRAL-SAR) algorithm: From statistical and algebraic fundamentals to quantum consequences, In *Quantum frontiers of atoms and molecules*, M.V. Putz (Ed.), *Series chemistry research and applications*, NOVA Science Publishers, Inc., New York, 2010, Chapter 21.

16. Hypercube, Inc. (2002) HyperChem 7.01 [Program package].

17. Putz, M.V.; Lacrămă, A.M.; and Ostafe, V. Full analytic progress curves of the enzymic reactions *in vitro. Int. J. Mol. Sci.* **2006**, *7*, 469–484.

18. Putz, M.V. and Lacrămă, A.M. Enzymatic control of the bio-inspired nanomaterials at the spectroscopic level. *J. Optoel. Adv. Mat.* **2007**, *9(8)*, 2529–2534.

19. Putz, M.V.; Lacrămă, A.M.; and Ostafe, V. Introducing logistic enzyme kinetics. *J. Optoel. Adv. Mat.* **2007**, *9(9)*, 2910–2916.

6

1. Pernak, J. and Chwala, P. Synthesis and anti-microbial activities of choline-like quaternary ammonium chlorides. *Eur. J. Med. Chem.* **2003**, *38*, 1035–1042.

2. Bernot, R.J.; Brueseke, M.A.; Evans-White, M.A.; and Lamberti, G.A. Acute and chronic toxicity of imidazolium-based ionic liquids on *Daphnia Magna*. *Environ. Toxicol. Chem.* **2005**, *24*, 87–92.

3. Sheldon, R.A. Green solvents for sustainable organic synthesis: state of the art. *Green Chem.* **2005**, *7*, 267–278.

4. Docherty, K.M. and Kulpa, C.F.Jr. Toxicity and antimicrobial activity of imidazolium and pyridinium ionic liquids. *Green Chem.* **2005**, *7*, 185–189.

5. Freemantle, M. New frontiers for ionic liquids. *Chem. Eng. News* **2007**, *1*, 23–26.

6. Anastas, P.T. and Warner, J.C. *Green Chemistry Theory and Practice*. Oxford University Press, New York, **1998**.

7. Jastorff, B.; Molter, K.; Behrend, P.; Bottin-Weber, U.; Filser, J.; Heimers, A.; Ondurschka, B.; Ranke, J.; Scaefer, M.; Schroder, H.; Stark, A.; Stepnowski, P.; Stock, F.; Stormann, R.; Stolte, S.; Welz-Biermann, U.; Ziegert, S.; and Thoming, J. Progress in evaluation of risk potential of ionic liquids—basis for an eco-design of sustainable products. *Green Chem.* **2005**, *7*, 362–372.

8. Wells, A.S. and Coombe, V.T. On the freshwater ecotoxicity and biodegradation properties of some common ionic liquids. *Org. Process Res. Dev.* **2006**, *10*, 794–798.

9. Garcia, M.T.; Gathergood, N. and Scammells, P.J. Biodegradable ionic liquids. Part II. Effect of the anion and toxicology. *Green Chem.* **2005**, *7*, 9–14.

10. Jain, D.; Kumar, A.; Chauhan, S.; and Chauhan, S.M.S. Chemical and biochemical transformation in ionic liquids. *Tetrahedron* **2005**, *61*, 1015–1060.

11. Dupont, J. and Suarez, P.A.Z. Physico-chemical processes in imidazolium ionic liquids. *Phys. Chem. Chem. Phys.* **2006**, *8*, 2441–2452.

12. Hansen, J.P. and McDonald, I.R. *Theory of simple liquids,* 2nd ed. Academic Press, London, 1986.

13. Scammells, P.J.; Scott, J.L. and Singer, R.D. Ionic liquids: The neglected issues. *Aust. J. Chem.* **2005**, *58*, 155–169.

14. Hunt, P.A.; Gould, I.R. and Kirchner, B. The structure of imidazolium-based ionic liquids: insights from ion-pair interactions. *Aust. J. Chem.* **2007**, *60*, 9–14.

15. Hunt, P.A. and Kirchner, B.; Welton, T. Characterizing the electronic structure of ionic liquids: An examination of the 1-butyl-3-ethylimidazolium chloride ion pair. *Chem. Eur. J.* **2006**, *12*, 6762–6775.

16. Hunt, P.A. and Gould, I.R. Structural characterization of the 1-butyl-3-methylimidazolium chloride ion pair using ab initio methods. *J. Phys. Chem. A.* **2006**, *110*, 2269–2282.

17. Hunt, P.A. The Simulation of imidazolium-based ionic liquids. *Mol. Simul.* **2006**, *32*, 1–10.

18. Putz, M.V. and Lacrămă, A.M. Introducing spectral structure activity relationship (SPECTRAL-SAR) analysis. Application to ecotoxicology. *Int. J. Mol. Sci.* **2007**, *8*, 363–391.

19. Jastorff, B.; Stormann, R.; Ranke, J.; Molter, K.; Stock, F.; Oberheitmann, B.; Hoffmann, W.; Hoffmann, J.; Nuchter, M.; Ondruschka, B.; and Filser, J. How hazardous are ionic liquids? Structure—activity relationship and biologic testing as important elements for sustainability evaluation. *Green Chem.* **2003**, *5*, 136–142.

20. Pernak, J.; Sobaszkiewicz, K.;and Mirska, I. Antimicrobial activities of ionic liquids. *Green Chem.* **2003**, *5*, 52–56.

21. Lacrămă, A.M.; Putz, M.V.; and Ostafe V. Designing a spectral structure-activity ecotoxico-logistical battery. In *Advances in quantum chemical bonding structures*, M.V. Putz (Ed.). Research Signpost, Kerala, India, 2008, Chapter 16, pp. 389–419.

22. National Toxicology Program (NTP) and National Institute of Environmental Health Sciences (NIEHS). *Review of toxicological literature for ionic liquids*, **2004**, Prepared By Integrated Laboratory Systems Inc., Research Triangle Park.

23. Bernot, R.J.; Kennedy, E.E.; and Lamberti, G.A. Effects of ionic liquids on the survival, movement, and feeding behavior of the freshwater snail. *Physa. Acuta. Environ. Toxicol. Chem.* **2005**, *24*, 1759–1765.

24. Couling, D.J.; Bernot, A.R.; Docherty, K.M.; Dixon, J.K.; and Maginn, E.J. Assessing the factors responsible for ionic liquid toxicity to aquatic organisms via quantitative structure—property relationship modeling. *Green Chem.* **2006**, *8*, 82–90.

25. Hypercube, Inc. 2002, HyperChem 7.01, Program package.

26. Stepnowski, P.; Skladanowski, A.C.; Ludwiczak, A.; and Laczynska, E. Evaluating the cytotoxicity of ionic liquids using human cell line hela. *Hum. Exp. Toxicol.* **2004**, *23*, 513–517.

27. Hansch, C. and Leo, A. *Exploring QSAR*, ACS Professional Reference Book. ACS, Washington, DC, 1995.

28. Swatloski, R.P.; Holbrey, J.D.; and Rogers, R.D. Ionic liquids are not always green: hydrolysis of 1-butyl-3-methylimidazolium hexafluorophosphate. *Green Chem.* **2003**, *5*, 361–363.

29. Kamrin, M.A. *Pesticide profiles: Toxicity, environmental impact, and fate*, Lewis Publishers, Boca Raton, Florida, 1997.

30. Docherty, K.M.; Hebbeler, S.Z.; and Kulpa, C.F.Jr. An assessment of ionic liquid mutagenicity using the ames test. *Green Chem.* **2006**, *8*, 560–567.

31. Stock, F.; Hoffmann, J.; Ranke, J.; Stormann, R.; Ondruschka, B.; and Jastorff, B. Effects of ionic liquids on the acetylcholinesterase— A structure-activity relationship consideration. *Green Chem.* **2004**, *6*, 286–290.

32. Raves, M.L.; Harel, M.; Pang, Y.P.; Silman, I.; Kozikowski, A.P.; and Sussman, J.L. 3D structure of acetylcholinesterase complexed with the nootropic alkaloid, (-)-huperzine A. *Nat. Struct. Biol.* **1997**, *4*, 57–63.

33. Skladanowski, A.C.; Stepnowski, P.; Kleszczynski, K.; and Dmochowska, B.

AMP deaminase in vitro inhibition by xenobiotics. A potential molecular method for risk assessment of synthetic nitro- and polycyclic musks, imidazolium ionic liquids and *n*-glucopyranosyl ammonium salts. *Environ. Toxicol. Phar.* **2005**, *19*, 291–296.

34. Ranke, J.; Molter Stock, F.; Bottin-Weber, U.; Poczobutt, J.; Hoffmann, J.; Ondruschka, B.; Filser, J.; and Jastorff, B. Biological effects of imidazolium ionic liquids with varying chain lengths in acute *Vibrio Fischeri* and wst-1 cell viability assays. *Ecotoxicol. Environ. Saf.* **2004**, *58*, 396–404.

35. Standard Test Method for Assessing the Microbial Detoxification of Chemically Contaminated Water and Soil Using a Toxicity Test with a Luminescent Marine Bacterium. *ASTM Designation: D 5660–96*, **1996**.

36. Kaiser, K.L.E. and Palabrica, V.S. *Photobacterium phosphoreum*, Toxicity data index. *Water Poll. Res. J. Can.* **1991**, *26*, 361–431.

37. McQueen, D.J.; Post, J.R.; Mills, E.L.; and Fish, C.J. Trophic relationships in freshwater pelagic eco-systems. *Can. J. Fish. Aquat. Sci.* **1986**, *43*, 1571–1581.

38. Swatloski, R.P.; Holbrey, J.D.; Memon, S.B.; Caldwell, G.A.; Caldwell, K.A.; and Rogers, R.D. Using *Caenorhabditis elegans* to probe toxicity of 1-alkyl-3-methylimidazolium chloride based ionic liquids. *Chem. Commun.* **2004**, 668–669.

39. Wong, P.T. and Couture, P. Toxicity screening using phytoplankton. In B.J. Dutka, and G. Bitton (Eds.). *Toxicity testing using microorganisms*, CRC Press, Boca Raton, Fl, 1986, Vol. II, pp. 79–100.

40. Latala, A.; Stepnowski, P.; Nedzi, M.; and Mrozik, W. Marine toxicity assessment of imidazolium ionic liquids: Acute effects on the baltic algae *Oocystis submarina* and *Cyclotella meneghiniana*. *Aquat. Toxicol.* **2005**, *73*, 91–98.

41. EN Water Quality—Fresh water algal growth-inhibition test with *Scenedesmus Subspicatus* and *Selenastrum Capricornutum* (ISO 8692:1993). *European Committee for Standardization*, **1993**, Brussels.

42. EN Water Quality—Marine Algal growth-inhibition test with *Skeletonema Costatum* and *Phaeodactylum Tricornutum* (ISO

10253:1995). *European Committee for Standardization*, **1995**, Brussels.

43. Cross, J. *Introduction to cationic surfactants: Analytical and biological evaluation*, 3rd ed. Marcel Dekker Inc., New York, 1994.

44. Pretti, C.; Chappe, C.; Pieraccini, D.; Gregori, M.; Abramo, F.; Monni, G.; and Intorre, L. Acute toxicity of ionic liquids to the zebrafish (*Danio rerio*). *Green Chem.* **2006**, *8*, 238–240.

45. Newman, M. and Dixon, P. Ecologically meaningful estimates of lethal effect in individuals. In *Ecotoxicology: A hierarchical treatment*, M.C. Newman and C.H. Jagoe (Eds.). Lewis, Boca Raton, U.S.A., 1996, pp. 225–253.

46. Gathergood, N.; Garcia, M.T.; and Scammells, P.J. Biodegradable ionic liquids: Part I. Concept, preliminary targets and evaluation. *Green Chem.* **2004**, *6*, 166–175.

47. Gathergood, N.; Scammells, P.J.; and Garcia, M.T. Biodegradable ionic liquids. Part III. The first readily biodegradable ionic liquids. *Green Chem.* **2006**, *8*, 156–160.

48. Ropel, R.; Belveze L.S.; Aki, S.N.V.K.; Stadtherr, M.A.; and Brennecke, J.F. Octanol-water partition coefficients of imidazolium-based ionic liquids. *Green Chem.* **2005**, *7*, 83–90.

49. Stolte, S.; Arning, J.; Bottin-Weber, U.; Matzke, M.; Stock, F.; Thiele, K.; Uerdingen, M.; Welz-Biermann, U.; Jastorff, B. ; and Ranke, J. Anion effects on the cytotoxicity of ionic liquids. *Green Chem.* **2006**, *8*, 621–629.

50. Shugart, L. Molecular markers to toxic agents. In *Ecotoxicology: A hierarchical treatment.* M.C. Newman and C.H. Jagoe (Eds.). Lewis, Boca Raton, Fl, **1996**, pp. 133–161.

51. Stiefl, N. and Bauman, K. Structure-based validation of the 3D-QSAR technique MaP. *J. Chem. Inf. Mod.* **2005**, *45*, 739–749.

7

1. Putz, M.V. A spectral approach of the molecular structure—Biological activity relationship Part I. The general algorithm. *Ann. West Univ. Timişoara, Ser. Chem.* **2006**, *15*, 159–166.

2. Putz, M.V. and Lacrămă, A.M. A spectral approach of the molecular structure—Biological activity relationship Part II. The enzymatic activity. *Ann. West Univ. Timişoara, Ser. Chem.* **2006**, *15*, 167–176.

3. Putz, M.V. and Lacrămă, A.M. Introducing spectral structure activity relationship (SPECTRAL-SAR) analysis. Application to ecotoxicology. *Int. J. Mol. Sci.* **2007**, *8*, 363–391.

4. Lacrămă, A.M.; Putz, M.V.; and Ostafe, V. A SPECTRAL-SAR model for the anionic-cationic interaction in ionic liquids: Application to *Vibrio fischeri* ecotoxicity. *Int. J. Mol. Sci.* **2007**, *8*, 842–863.

5. Lacrămă, A.M.; Putz, M.V.; and Ostafe, V. Designing a spectral structure-activity ecotoxico-logistical battery. In *Advances in quantum chemical bonding structures*, M.V. Putz (Ed.). Research Signpost, Kerala, India, 2007, Chapter 16, pp. 389–419.

6. *D. magna* is biological classified as: Kingdom *Animalia* > Phylum *Arthropoda* > Subphylum *Crustacea* > Class *Branchiopoda* > Order *Cladocera* > Family *Daphniidae* > Genus *Daphnia* > Subgenus *Ctenodaphnia* > Species *D. magna.* http://en.wikipedia.org (27.06.2007)

7. Verrhiest, G.; Clement, B; and Blake, G. Single and combined effects of sediment-associated PAHs on three species of freshwater macroinvertebrates. *Ecotoxicol.* **2001**, *10*, 363–372.

8. Couling, D.J.; Bernot, A.R.; Docherty, K.M.; Dixon, J.K.; and Maginn, E.J. Assessing the factors responsible for ionic liquid toxicity to aquatic organisms via quantitative structure—Property relationship modeling. *Green Chem.* **2006**, *8*, 82–90.

9. Connon, R.; Dewhurst, R.E.; Crane, M.; and Callaghan, A. Haem peroxidase activity in *Daphnia magna*: A biomarker for a sublethal toxicity assessment of kerosene-contaminated groundwater. *Ecotoxicol.* **2003**, *12*, 387–395.

10. Smith, D.G. *Pennak's freshwater invertebrates of the United States: Porifera to crustacea*, 4th ed. Wiley, New York, 2001.

11. Diamantino, T.C.; Almeida, E.; Soares, A.M.V.M.; and Guilhermino, L. Lactate dehydrogenase activity as an effect criterion in

toxicity tests with *Daphnia magna* straus. *Chemosphere* **2001**, *45*, 553–560.

12. EEC Directive 92/32/EEC. Seventh amendment of Directive 67/548/EEC, Annex. V. Part C: Methods for the determination of ecotoxicity. C2: Acute toxicity for Daphnia. J.O. No. 154, 5/6/**1992**.

13. United States Environmental Protection Agency. Methods for measuring the acute toxicity of effluents and receiving waters to freshwater and marine organisms, EPA 600/4-90/027F, Cincinnati, OH, **1993**.

14. International Organisation for Standardisation. Water quality determination of the inhibition of the mobility of *Daphnia magna* straus (Cladocera, Crustacea), ISO Geneva, 6341, **1982**.

15. Seco, J.; Fernández-Pereira, C.; and Vale, J. A study of the leachate toxicity of metal-containing solid waste using *Daphnia magna*. *Ecotoxicol. Environ. Saf.* **2003**, *56*, 339–350.

16. Hirano, M.; Ishibashi, H.; Matsumura, N.; Nagao, Y.; Watanabe, N.; Watanabe, A.; Onikura, N.; Kishi, K.; and Arizono, K. Acute toxicity responses of two crustaceans, *Americamysis bahia* and *Daphnia magna*, to endocrine disrupters *J. Health Sci.* **2004**, *50*, 97–100.

17. Bernot, R.J.; Brueseke, M.A.; Evans-White, M.A.; and Lamberti, G.A. Acute and chronic toxicity of imidazolium-based ionic liquids on *Daphnia Magna*. *Environ. Toxicol. Chem.* **2005**, 24, 87–92.

18. Hypercube, Inc. 2002, HyperChem 7.01, Program package.

8

1. The electric eel (*Electrophorus electricus*) is classified as: Kingdom: Animalia > Phylum: Chordata > Class: Osteichthyes > Order: Gymnotiformes > Family: Gymnotidae > Genus: *Electrophorus* > Species: *E. electricus*; it is a species of fish which is capable of generating powerful electric shocks, used for both hunting and self-defense. It is an apex predator in its South American range; despite its name it is not an eel at all but rather a knife-fish (http://en.wikipedia.org/wiki/Electric_eel, inspected on 07.08.2008).

2. Gotter, A.L.; Kaetzel, M.A.; and Dedman, J.R. *Electrophorus electricus* as a model system for the study of membrane excitability. *Comp. Biochem. Physiol. Part A: Mol. Integ. Physiol.* **1998**, *119*(1), 225–241.

3. Stock, F.; Hoffmann, J.; Ranke, J.; Stormann, R.; Ondruschka, B.; and Jastorff, B. Effects of ionic liquids on the acetylcholinesterase––A structure-activity relationship consideration. *Green Chem.* **2004**, *6*, 286–290.

4. Stolte, S.; Arning, J.; Bottin-Weber, U.; Muller, A.; Pitner, W.R.; Welz-Biermann, U.; Jastorff, B.; and Ranke, J. Effects on different head groups and functionalised side chains on the cytotoxicity of ionic liquids. *Green Chem.* **2007**, *9*, 760–767.

5. Pernak, J. and Chwala, P. Synthesis and anti-microbial activities of choline-like quaternary ammonium chlorides. *Eur. J. Med. Chem.* **2003**, *38*, 1035–1042.

6. Freemantle, M. New frontiers for ionic liquids. *Chem. Eng. News* **2007**, *1*, 23–26.

7. Jastorff, B.; Stormann, R.; Ranke, J.; Molter, K.; Stock, F.; Oberheitmann, B.; Hoffmann, J.; Nuchter, M.; Ondruschka, B.; and Filser, J. How hazardous are ionic liquids? Structure-activity relationship and biological testing as important elements for sustainability evaluation. *Green Chem.* **2003**, *5*, 136–142.

8. Sheldon, R.A. Green solvents for sustainable organic synthesis: State of the art. *Green Chem.* **2005**, *7*, 267–278.

9. (a) Lacrămă, A.M.; Putz, M.V.; and Ostafe, V. A SPECTRAL-SAR model for the anionic-cationic interaction in ionic liquids: Application to *Vibrio fischeri* Ecotoxicity. *Int. J. Mol. Sci.* **2007**, *8*, 842–863.

  (b) Putz, M.V.; Lacrămă, A.M.; and Ostafe, V. SPECTRAL-SAR ecotoxicology of ionic liquids. The *Daphnia magna* case. *Res. Lett. Ecol.* **2007**, Article ID12813/5 pages, DOI: 10.1155/2007/12813.

10. (a) Putz, M.V. A Spectral approach of the molecular structure—Biological activity relationship. Part I. The general algorithm. *Ann. West Univ. Timişoara, Ser. Chem.* **2006**, *15*, 159–166.

  (b) Putz, M.V. and Lacrămă, A.M. A spectral approach of the molecular structure—Biological activity relationship. Part II.

The enzymatic activity. *Ann. West Univ. Timişoara, Ser. Chem.* **2006**, *15*, 167–176.

(c) Putz, M.V. and Lacrămă, A.M. Introducing spectral structure activity relationship (SPECTRAL-SAR) analysis. Application to ecotoxicology. *Int. J. Mol. Sci.* **2007**, *8*, 363–391.

(d) Putz, M.V.; Duda-Seiman, C.; Duda-Seiman, D.M.; and Putz, A.M. Turning SPECTRAL-SAR into 3D-QSAR analysis. Application on H^+K^+-ATPase inhibitory activity. *Int. J. Chem. Model.* **2008**, *1*, 45–62.

(e) Lacrămă, A.M.; Putz, M.V.; and Ostafe, V. Designing a spectral structure-activity ecotoxico-logistical battery. In *Advances in quantum chemical bonding structures*, M.V. Putz (Ed.). Transworld Research Network, Kerala, India, 2008, Chapter 16, pp. 389–419.

(f) Putz, M.V. and Putz (Lacrămă), A.M. SPECTRAL-SAR: Old wine in new bottle. *Studia Universitatis Babeş-Bolyai Chemia* **2008**, *53(2)*, 73–81.

11. Hypercube, Inc. (2002) HyperChem 7.01 [Program package, Semiempirical, AM1, Polak-Ribier optimization procedure].

12. Escher, B. Molecular mechanisms in aquatic ecotoxicology: Specific and non-specific membrane toxicity. ETH Zurich. *Habilitationsschrift.* 2001.

13. Johnson, I.; Hutchings, M.; Benstead, R.; Thain, J.; and Whitehouse, P. Bioassay selection, experimental design and quality control/assurance for use in effluent assessmnet and control. *Ecotoxicology* **2004**, *13*, 437–447.

14. Repetto, G.; Jos, A.; Hazen, M.J.; Molero, M.L.; del Peso, A.; Salguero, M.; del Castillo, P.; Rodryguez-Vicente, M.C.; and Repetto, M. A test battery for the ecotoxicological evaluation of pentachlorophenol. *Toxicology in Vitro* **2001**, *15*, 503–509.

15. CSTEE. Scientific committee on toxicity, ecotoxicity, and the environment: Opinion on the available scientific approaches to assess the potential effects and risk of chemicals on terrestrial ecosystems. Opinion expressed at the 19th CSTEE plenary meeting—Brussels, 9 November 2000.

16. Ahlers, J. Strategies for risk assessment of existing chemicals in soil. *J. Soils & Sediments* **2001**, *1*, 168–174.

17. Ahlers, J. and Martin, S. Global Soils: EU, risk assessment of chemicals in soil: Recent developments in the EU. *J. Soils & Sediments* **2003**, *3*, 240–241.

18. Hund-Rinke, K.; Kordel, W.; Hennecke, D.; Eisentrager, A.; and Heiden, St. Bioassays for the ecotoxicological and genotoxicological assessment of contaminated soils (Results of a Round Robin Test). Part I. Assessment of a possible groundwater contamination: Ecotoxicological and genotoxicological tests with aqueous soil extracts. *J. Soils & Sediments* **2002**, *2*, 43–50.

19. Hund-Rinke, K.; Kordel, W.; Hennecke, D.; Achazi, R.; Warnecke, D.; Wilke, B.-M.; Winkel, B.; and Heiden, St. Bioassays for the ecotoxicological and genotoxicological assessment of contaminated soils (Results of a Round Robin Test). Part II. Assessment of the habitat function of soils: Tests with soil microflora and fauna. *J. Soils & Sediments* **2002**, *2*, 83–90.

20. Achazi, R.K. Invertebrates in risk assessment development of a test battery and of short term biotests for ecological risk assessment of soil. *J. Soils & Sediments* **2002**, *2*, 174–178.

21. Jos, A.; Repetto, G.; Rios, J.C.; Hazen, N.; Molero, M.L.; del Peso, A.; Salguero, M.; Fernandez-Freire, P.; Perez-Martin, J.M.; and Camen, A. Ecotoxicological evaluation of carbamazepine using six different model systems with eighteen endpoints. *Toxicology in Vitro* **2003**, 17, 525–532.

22. Jos, A.; Repetto, G.; Rios, J.C.; Peso, A.D.; Salguero, M.; Camean, A.; Rios, J.C.; Hazen, M.J.; Molero, M.L.; Fernandez-Freire, P.; Perez-Martin, J.M.; and Labrador, V. Ecotoxicological evaluation of the additive butylated hydroxyanisole using a battery with six model systems and eighteen endpoints. *Aquatic Toxicology* **2005**, 71, 183–192.

23. Escher, B.I.; Behra, R.; Eggen, R.I.L.; and Fent, K. Molecular mechanisms in ecotoxicology: An interplay between environmental chemistry and biology. *Chimia* **1997**, *51*, 915–921.

24. Mincea, M.; Lacrămă, A.M.; and Ostafe, V. Use of bovine liver alkaline phosphatase for testing at molecular level the toxicity of chemical compounds. *Ann. West Univ. Timişoara, Ser. Chem.* **2004**, 13, 87–98.

25. Mincea, M.; Lacrămă, A.M.; Stoian, C.; Baicu, I.; Nemes, N.; Popet, L; and Ostafe, V. Multienzimatic test battery-a model for testing at molecular level the toxicity of chemical compounds. In *Sustainability for humanity and environment in the extended connection field science-economy-policy.* Polytechnic Publishing House, Timişoara, 2005, Vol. II, pp. 211–214, ISBN 973-625-206-X.

26. Schnell, S. and Maini, P.K. A century of enzyme kinetics: Reliability of the K_M and v_{max} estimates. *Comm. on Theor. Biol.* **2003**, 8, 169–187.

27. Lacrămă, A.M.; Putz, M.V.; and Ostafe, V. New enzymatic kinetic relating Michaelis–Menten mechanisms. *Ann. West Univ. Timişoara, Ser. Chem.* **2005**, 14, 179–190.

28. Putz, M.V.; Lacrămă, A.M.; and Ostafe, V. Full time course analysis for reversible enzyme kinetics. *Proceedings of the VIIIth International Symposium* YOUNG PEOPLE AND MULTIDISCIPLINARY RESEARCH Romania—Serbia—Hungary, 11–12 May 2006, Timişoara, pp. 642–649, ISBN (10) 973-8359-39-2 ; ISBN (13) 978-973-8359-39-0, Welding Publishing House.

29. Seddon, K.R. Ionic liquids. *Green Chem.* **2002**, 4, G25–26.

30. Sheldon, R. Catalytic reactions in ionic liquids. *Chem. Comm. (Cambridge)* **2001**, 23, 2399–2407.

31. Anastas, P.T. and Warner, J.C. *Green chemistry, theory and practice.* Oxford University Press, New York, 1998.

32. Jastorff, B.; Stormann, R.; and Wolke, U. Struktur-Wirkungs-Denken in der Chemie. Universitatsverlag Aschenbeck and Isensee, Bremen, Oldenburg, 2004.

33. Jastorff, B.; Molter, K.; Behind, P.; Bottin-Weber, U.; Filser, J.; Heimers, A.; Ondruschka, B.; Ranke, J.; Schaefer, M.; Schroder, H.; Stark, A.; Stepnowski, P.; Stock, F.; Stormann, R.; Stolte, S.; Welz-Biermannn, U.; Ziegerta, S.; and Thominga, J. Progress in evaluation of risk potential of ionic liquids—Basis for an eco-design of sustainable products. *Green Chem.* **2005**, 7, 362–372.

34. Ranke, J.; Molter, K.; Stock, F.; Bottin-Weber, U.; Poczobutt, J.; Hoffmann, J.; Ondruschka, B.; Filser, J.; and Jastorff, B. Biological effects of imidazolium ionic liquids with varying chain lengths in acute *Vibrio fischeri* and WST-1 cell viability assays *Ecotoxicol. Environ. Safety* **2004**, 58, 396–404.

9

1. Anderson, T.W. *An introduction to multivariate statistical methods.* Wiley, New York, USA, 1958.

2. Draper, N.R. and Smith, H. *Applied regression analysis.* Wiley, New York, USA, 1966.

3. Shorter, J. *Correlation analysis in organic chemistry: An introduction to linear free energy relationships.* Oxford University Press, London, UK, 1973.

4. Box, G.E.P.; Hunter, W.G.; and Hunter, J.S. *Statistics for experimenters.* John-Wiley, New York, USA, 1978.

5. Green, J.R. and Margerison, D. *Statistical treatment of experimental data.* Elsevier, New York, USA, 1978.

6. Topliss, J. *Quantitative structure-activity relationships of drugs.* Academic Press, New York, USA, 1983.

7. Seyfel, J.K. *QSAR and strategies in the design of bioactive compounds.* VCH Weinheim, New York, USA, 1985.

8. Chatterjee, S.; Hadi, A.S.; and Price, B. *Regression analysis by examples,* 3rd ed. John-Wiley, New York, USA, 2000.

9. European Commission. Regulation (EC) No. 1907/2006 of the European Parliament and of the Council of 18 Dec. 2006 concerning the registration, evaluation, authorisation, and restriction of chemicals (REACH), establishing a European Chemicals Agency, amending directive 1999/45/EC and repealing Council Regulation (EC) No. 1488/94 as well as Council Directive 76/769/EEC and commission directives 91/155/EEC, 93/67/EEC, 93/105/EC and 2000/21/EC. *Off. J. Eur. Union, L 396/1 of 30.12.2006.* Office for Official Publication of the

European Communities (OPOCE), Luxembourg, 2006.

10. European Commission. Directive 2006/121/ EC of the European Parliament and of the Council of 18 Dec. 2006 amending Council Directive 67/548/EEC on the approximation of laws, regulations and administrative provisions relating to the classification, packaging and labelling of dangerous substances in order to adapt it to Regulation (EC) No. 1907/2006 concerning the registration, evaluation, authorisation and restriction of chemicals (REACH) and establishing a European chemicals agency. *Off. J. Eur. Union, L 396/850 of 30.12.2006.* Office for Official Publication of the European Communities (OPOCE), Luxembourg, 2006.

11. OECD, Report on the regulatory uses and applications in OECD member countries of (quantitative) structure-activity relationship [(Q)SAR] models in the assessment of new and existing chemicals. Organization of Economic Cooperation and Development: Paris, France, 2006; Available online: http:// www.oecd.org/, accessed January 2009.

12. OECD, Guidance document on the validation of (quantitative) structure-activity relationship [(Q)SAR] models. OECD series on testing and assessment No. 69. ENV/JM/ MONO (2007) 2. Organization for Economic Cooperation and Development: Paris, France, 2007; Available online: http://www. oecd.org/, accessed January 2009.

13. Worth, A.P.; Bassan, A.; Gallegos Saliner, A.; Netzeva, T.I.; Patlewicz, G.; Pavan, M.; Tsakovska, I.; and Vracko, M. The characterization of quantitative structure-activity relationships: Preliminary guidance. European Commission—Joint Research Centre, Ispra, Italy, 2005. Available online: http:// ecb.jrc.it/qsar/publications/, accessed January 2009.

14. Worth, A.P.; Bassan, A.; Fabjan, E.; Gallegos Saliner, A.; Netzeva, T.I.; Patlewicz, G.; Pavan, M.; and Tsakovska, I. The characterization of quantitative structure-activity relationships: Preliminary guidance. European Commission—Joint Research Centre, Ispra, Italy, 2005. Available online: http:// ecb.jrc.it/qsar/publications/, accessed January 2009.

15. Benigni, R.; Bossa, C.; Netzeva, T.I.; and Worth, A.P. Collection and evaluation of [(Q)SAR] models for mutagenicity and carcinogenicity. European Commission-Joint Research Centre, Ispra, Italy, 2007. Available online: http://ecb.jrc.it/qsar/publications/, accessed January 2009.

16. So, S.S. and Karpuls, M. Evolutionary optimisation in quantitative structure-activity relationship: An application of genetic neural network. *J. Med. Chem.* **1996**, *39*, 1521–1530.

17. Kubinyi, H. Evolutionary variable selection in regression and PLS analysis. *J. Chemometr.* **1996**, *10*, 119–133.

18. Teko, I.V.; Alessandro, V.A.E.P.; and Livingston, D.J. Neutral network studies. 2. Variable selection. *J. Chem. Inf. Comput. Sci.* **1996**, *36*, 794–803.

19. Kubinyi, H. Variable selection in QSAR studies. 1. An evolutionary algorithm. *Quant. Struct.-Act. Relat.* **1994**, *13*, 285–294.

20. Haegawa, K.; Kimura, T.; and Fanatsu, K. GA strategy for variable selection in QSAR Studies: Enhancement of comparative molecular binding energy analysis by GA-based PLS method. *Quant. Struct.-Act. Relat.* **1999**, *18*, 262–272.

21. Zheng, W. and Tropsha, A. Novel variable selection quantitative structure-property relationship approach based on the k-nearest neighbour principle. *J. Chem. Inf. Comput. Sci.* **2000**, *40*, 185–194.

22. Lucic, B. and Trinajstic, N. Multivariate regression outperforms several robust architectures of neural networks in QSAR modelling. *J. Chem. Inf. Comput. Sci.* **1999**, *39*, 121–132.

23. Duchowicz, P.R. and Castro, E.A. *The order theory in QSPR-QSAR studies*; Mathematical Chemistry Monographs. University of Kragujevac, Kragujevac, Serbia, 2008.

24. Zhao, V.H.; Cronin, M.T.D.; and Dearden, J.C. Quantitative structure-activity relationships of chemicals acting by non-polar narcosis—Theoretical considerations. *Quant. Struct.-Act. Relat.* **1998**, *17*, 131–138.

25. Pavan, M.; Netzeva, T.; and Worth, A.P. Review of literature based quantitative structure-activity relationship models for

bioconcentration. *QSAR Comb. Sci.* **2008**, *27*, 21–31.

26. Pavan, M. and Worth, A.P. Review of estimation models for biodegradation. *QSAR Comb. Sci.* **2008**, *27*, 32–40.

27. Tsakovska, I.; Lessigiarska, I.; Netzeva, T.; and Worth, A.P. A mini review of mammalian toxicity (Q)SAR models. *QSAR Comb. Sci.* **2008**, *27*, 41–48.

28. Gallegos Saliner, A.; Patlewicz, G.; and Worth, A.P. A review of (Q)SAR models for skin and eye irritation and corrosion. *QSAR Comb. Sci.* **2008**, *27*, 49–59.

29. Patlewicz, G.; Aptula, A.; Roberts, D.W. and Uriarte, E. A mini-review of available skin sensitization (Q)SARs/Expert systems. *QSAR Comb. Sci.* **2008**, *27*, 60–76.

30. Netzeva, T.; Pavan, M.; and Worth, A.P. Review of (quantitative) structure-activity relationship for acute aquatic toxicity. *QSAR Comb. Sci.* **2008**, *27*, 77–90.

31. Cronin, M.T.D. and Worth, A.P. (Q)SARs for predicting effects relating to reproductive toxicity. *QSAR Comb. Sci.* **2008**, *27*, 91–100.

32. Putz, M.V. A spectral approach of the molecular structure—biological activity relationship. Part I. The general algorithm. *Ann. West Univ. Timişoara Ser. Chem.* **2006**, *15*, 159–166.

33. Putz, M.V. and Lacrămă, A.M. A spectral approach of the molecular structure—biological activity relationship. Part II. The enzymatic activity. *Ann. West Univ. Timişoara Ser. Chem.* **2006**, *15*, 167–176.

34. Putz, M.V. and Lacrămă, A.M. Introducing spectral structure activity relationship (SPECTRAL-SAR) analysis. Application to ecotoxicology. *Int. J. Mol. Sci.* **2007**, *8*, 363–391.

35. Lacrămă, A.M.; Putz, M.V.; and Ostafe, V. A SPECTRAL-SAR model for the anionic-cationic interaction in ionic liquids: Application to *Vibrio fischeri* ecotoxicity. *Int. J. Mol. Sci.* **2007**, *8*, 842–863.

36. Putz, M.V.; Lacrămă, A.M.; and Ostafe V. SPECTRAL-SAR ecotoxicology of ionic liquids. The *Daphnia magna* case. *Int. J. Ecol. (Res. Lett. Ecol.)* **2007**, Article ID12813/5 pages, DOI: 10.1155/2007/12813.

37. Putz, M.V.; Duda-Seiman, C.; Duda-Seiman, D.M.; and Putz A.M. Turning SPECTRAL-SAR into 3D-QSAR analysis. Application on H^+K^+-ATPase inhibitory activity. *Int. J. Chem. Model.* **2008**, *1*, 45–62.

38. Lacrămă, A.M.; Putz, M.V.; and Ostafe, V. Designing a spectral structure-activity ecotoxico-logistical battery. In *Advances in quantum chemical bonding structures*, M.V. Putz (Ed.). Transworld Research Network, Kerala, India, Chapter 16, pp. 389–419, 2008.

39. Putz, M.V. and Putz (Lacrămă) A.M. SPECTRAL-SAR: Old wine in new bottle. *Studia Universitatis Babeş-Bolyai Chemia*, **2008**, *53*, 73–81.

40. Putz, M.V.; Putz, A.M.; Ostafe, V.; and Chiriac A. Application of spectral-structure activity relationship (SPECTRAL-SAR) method to ecotoxicology of some ionic liquids at the molecular level using acethyl-colinesterase. *Int. J. Chem. Model.* **2010**, *2*, 85–96.

41. Steiger, J.H. and Schonemann, P.H. A history of factor indeterminacy. In *Theory construction and data analysis in the behavioural science*, S. Shye (Ed.). Jossey-Bass Publishers, San Francisco, CA, USA, 1978.

42. Spearman, C. *The abilities of man.* MacMillan, London, UK, 1927.

43. Wilson, E.B. Review of the abilities of man, their nature and measurement, by Spearman, C. *Science* **1928**, *67*, 244–248.

44. Wilson, E.B. and Hilferty, M.M. The distribution of chi-square. In *Proc. Nat. Acad. Sci. USA* **1931**, *17*, 684.

45. Wilson, E.B. and Worcester, J. A note on factor analysis. *Psychometrika* **1939**, *4*, 133–148.

46. Topliss, J.G. and Costello, R.J. Chance correlation in structure-activity studies using multiple regression analysis. *J. Med. Chem.* **1972**, *15*, 1066–1068.

47. Topliss, J.G. and Edwards, R.P. Chance factors in studies of quantitative structure-activity relationships. *J. Med. Chem.* **1979**, *22*, 1238–1244.

48. Dittrich, W. and Reuter, M. *Classical and quantum dynamics: From classical paths to path integrals.* Springer-Verlag, Berlin, Germany, 1992.

49. Havsteen, B.H. The biochemistry and medical significance of the flavonoids. *Pharmacol. Ther.* **2002**, *96*, 67–202.

50. Middleton, E.Jr.; Kandaswami, C.; and Theoharides, T.C. The effects of plant flavonoids on mammalian cells: Implications for inflammation, heart disease, and cancer. *Pharmacol. Rev.* **2000**, *52*, 673–751.

51. Zhang, S.; Yang, X.; Coburn, R.A.; and Morris, M.E. Structure activity relationships and quantitative structure activity relationships for the flavonoid-mediated inhibition of breast cancer resistance protein. *Biochem. Pharmacol.* **2005**, *70*, 627–639.

52. Zhang, S; Yang, X; and Morris, M.E. Combined effects of multiple flavonoids on breast cancer resistance protein (ABCG2)-mediated transport. *Pharm. Res.* **2004**, *21*, 1263–1273.

53. Zhang, S; Yang, X; and Morris, M.E. Flavonoids are inhibitors of breast cancer resistance protein (ABCG2)-mediated transport. *Mol. Pharmacol.* **2004**, *65*, 1208–1216.

54. Sargent, J.M.; Williamson, C.J.; Maliepaard, M.; Elgie, A.W.; Scheper, R.J.; and Taylor, C.G. Breast cancer resistance protein expression and resistance to daunorubicin in blast cells from patients with acute myeloid leukaemia. *Br. J. Haematol.* **2001**, *115*, 257–262.

55. Hypercube, Inc. *HyperChem 7.01, Program package, Semiempirical, AM1, Polak-Ribiere optimization procedure*, 2002.

56. Hansch, C.A. A quantitative approach to biological-structure activity relationships. *Acta Chem. Res.* **1969**, *2*, 232–239.

57. Miller, J.N. and Miller, J.C. Statistics and Chemometrics for Analytical Chemistry, 4th ed. Pretience Hall, Harlow, England, 2000.

58. StatSoft, Inc. STATISTICA for Windows, Computer program and manual, 1995.

10

1. (a) Balaban, A.T.; Chiriac, A.; Motoc, I.; and Simon, Z. *Steric fit in quantitative structure-activity relations*, Lecture notes in chemistry Vol. 15. Springer-Verlag, Berlin, 1980.

    (b) Stone, M. and Jonathan, P. Statistical thinking and technique for QSAR and related studies. 1. Genaral theory. *J. Chemometrics* **1993**, *7*, 455–475.

    (c) Stone, M. and Jonathan, P. Statistical thinking and technique for QSAR and related studies. Part II: Specific methods. *J. Chemometrics* **1994**, *8*, 1–20.

    (d) Graovac, A.; Gutman, I.; and Trinajstić, N. *Topological approach to the chemistry of conjugated molecules*. Springer-Verlag: Berlin, 1977.

    (e) Todeschini, R. and Consonni, V. *Handbook of molecular descriptors*. Wiley-VCH, Weinheim, 2000.

2. Draper, N.R. and Smith, H. *Applied regression analysis*. Wiley, New York, 1981.

3. Dunn III, W.J. and Wold, S. Structure-activity analyzed by pattern recognition: The asymmetric case. *J. Med. Chem.* **1980**, *23*, 595–599.

4. (a) Mager, P.P. and Rothe, H. Obscure phenomena in statistical analysis of quantitative structure-activity relationships. Part 1: Multicollinearity of physicochemical descriptors. *Pharmazie* **1990**, *45*, 758–764.

    (b) Mager, P.P. Non-least-squares regression analysis applied to organic and medicinal chemistry. *Med. Res. Rev. (N.Y.)* **1994**, *14*, 533–588.

5. (a) Wold, S. and Dunn III, W.J. Multivariate quantitative structure-activity relationships (QSAR): Conditions for their applicability. *J. Chem. Inf. Comput. Sci.* **1983**, *23*, 6–13.

    (b) Box, G.E.; Hunter, W.G.; and Hunter, J.S. *Statistics for experiments*. Wiley, New York, 1978.

    (c) Toplis, J.G. and Costello, R.J. Chance correlation in structure-activity studies using multiple regression analysis. *J. Med. Chem.* **1972**, *15*, 1066–1069.

    (d) Toplis, J.G.and Edwards, R.P. Chance factors in studies of quantitative structure-activity relationships. *J. Med. Chem.* **1979**, *22*, 1238–1244.

6. (a) Mager, P.P.; Rothe, H.; Mager, H.; and Werner; H. In *QSAR in design of bioactive compounds*, M. Kuchar (Ed.). J.R. Prous Science Publishers, Barcelona, pp. 131–182, 1992.

(b) Mager, P.P. Diagnostics statistics in QSAR. *J. Chemometrics* **1995**, *9*, 211–221.

(c) Mager, P.P. A rigorous QSAR analysis? *J. Chemometrics* **1995**, *9*, 232–236.

7. Geisser, S. The predictive sample reuse method with applications. *J. Am. Stat. Assoc.* **1975**, *70*, 320–328.

8. (a) Allen, D.M. The relationship between variable selection and data augmentation and a method for prediction. *Technometrics* **1974**, *16*,125–127.

(b) Geisser, S. A prediction approach to the random effect model. *Biometria* **1974**, *61*, 101–107.

9. Wold, S.; Ruhe, A.; Wold, H.; and Dunn III, W.J. The collinearity problem in linear regression, the partial least squares (PLS) approach to generalized inverses. *SIAM J. Sci. Stat. Comput.* **1984**, *5*, 735–743.

10. Benigni, C. and Giuliani, A. What kind of statistics for QSAR research? *Quant. Struct.-Act. Relat.* **1991**, *10*, 99–100.

11. (a) Klein, D.J.; Randić, M.; Babić, D.; Lučić, B.; Nikolić, S.; and Trinajstić, N. Hierarchical orthogonalization of descriptors. *Int. J. Quantum Chem.* **1997**, *63*, 215–222.

(b) Soskić, M.; Plavsić, D.; and Trinajstić, N. 2-difluoromethylthio-4,6-bis(monoalkylamino)-1,3,5-triazines as inhibitors of Hill reaction: A QSAR study with orthogonalized descriptors. *J. Chem. Inf. Comput. Sci.* **1996**, *36*, 146–150.

(c) Soskić, M.; Plavsić, D.; and Trinajstić, N. Inhibition of the Hill reaction by 2-methylthio-4,6-bis(monoalkylamino)-1,3,5-triazines. *J. Mol. Struct. (Theochem)* **1997**, *394*, 57–65.

12. (a) Putz, M.V. and Lacrămă, A.M. Introducing spectral structure activity relationship (SPECTRAL-SAR) analysis. Application to ecotoxicology. *Int. J. Mol. Sci.* **2007**, *8*, 363.

(b) Lacrămă, A.M.; Putz, M.V.; and Ostafe, V.A. SPECTRAL-SAR model for the anionic-cationic interaction in ionic liquids: Application to *Vibrio fischeri* ecotoxicity. *Int. J. Mol. Sci.* **2007**, *8*, 842.

(c) Putz, M.V.; Lacrămă, A.M.; and Ostafe, V. SPECTRAL-SAR ecotoxicology of ionic liquids. The *Daphnia magna* case. *Int. J. Ecol. (Res. Lett. Ecol.)* **2007**, Article ID12813/5 pages, DOI: 10.1155/2007/12813;

(d) Lacrămă, A.M.; Putz, M.V.; and Ostafe, V. Designing a spectral structure-activity ecotoxico-logistical battery, In *Advances in quantum chemical bonding structures*, M.V. Putz (Ed.). Transworld Research Network, Kerala, India, Chapter 16, pp. 389–419, 2008.

(e) Putz, M.V.; Duda-Seiman, C.; Duda-Seiman, D.M.; and Putz, A.M. Turning SPECTRAL-SAR into 3D-QSAR analysis. application on H$^+$K$^+$-ATPase inhibitory activity. *Int. J. Chem. Model.* **2008**, *1*, 45–62;

(f) Putz, M.V.; Putz, A.M.; Lazea, M.; Ienciu, L.; and Chiriac, A. quantum-SAR extension of the SPECTRAL-SAR algorithm. Application to polyphenolic anticancer bioactivity. *Int. J. Mol. Sci.* **2009**, *10*, 1193–1214.

13. (a) Greim, H.; Bury, D.; Klimisch, H.J.; Oeben-Negele, M.; and Ziegler-Skylakakis, K. Toxicity of aliphatic amines: Structure-activity relationship. *Chemosphere* **1998**, *36*, 271–295.

(b) Devillers, J. and Balaban, A.T. (Eds.). *Topological indices and related descriptors in QSAR and QSPR*. Gordon and Breach Science Publishers, The Netherlands, pp. 279–306, 1999.

14. Gorrod, J.W. Differentiation of various types of biological oxidation of nitrogen in organic compounds. *Chem. Biol. Interact.* **1973**, *7*, 289–303.

15. BUA [Beratergremium für umweltrelevante Altstoffe (BUA) der GDCh (Hrsg.), id est: The Advisory Committee on Existing Chemicals of Environmental Relevance of the German Chemical Society] *Primäre Fettamine.* BUA-Stoffbericht 177, Germany, 1994.

16. *Corbin, D.R.; Schwarz, S.; and Sonnichsen, G.C. Methylamines synthesis: A review. Catalysis Today* **1997**, *37*, 71–102.

17. Giacobini, E. Piperidine: A new neuromodulator or a hypogenic substance? *Adv. Biochem. Psychopharm.* **1976**, *15*, 17–56.

18. van Gysel, B. and Musin, W. *Methylamines in Ullmann's encyclopedia of industrial chemistry.* Wiley-VCH Verlag, Weinheim, 2005.

19. Zhang, A.Q.; Mitchell, S.C.; and Smith, R.L. Dimethylamine formation in the rat

from various related amine precursors. *Food Chem. Toxicol.* **1998**, *36*, 923–927.

20. Eller, K.; Henkes, E.; Rossbacher, R.; and Höke, H. *Amines, aliphatic in Ullmann's encyclopedia of industrial chemistry*, Wiley-VCH, 2002.

21. International narcotics control board in accordance with the United Nations Convention against Illicit Traffic in Narcotic Drugs and Psychotropic Substances, containing the *List of precursors and chemicals frequently used in the ilicit manufacture of narcotic drugs and psychotropic substances under international control*. Vienna, Austria, 1988.

22. *Adams, R. and Brown, B. K.* Trimethylamine. *Org. Synth.* **1941**, *1*, 528–531.

23. Mitchell, S. C. and Smith, R. L. Trimethylaminuria: The fish malodor syndrome. *Drug metabolism and disposition* **2001**, *29*, 517–521.

24. BUA [Beratergremium für umweltrelevante Altstoffe (BUA) der GDCh (Hrsg.), id est: The Advisory Committee on Existing Chemicals of Environmental Relevance of the German Chemical Society]. *Tributylamin (N,N-Dibutylbutan-1-amin)*. BUA-Stoffbericht 23, Germany, 1988.

25. Lundh, T.; Boman, A.; and Åkesson, B. Skin absorption of the industrial catalyst dimethylamine in vitro in guinea pig and human skin, and of gaseous dimethylethylamine in human volunteers. *Int. Arch. Occup. Environ. Health* **1997**, *70*, 309–313.

26. Lundh, T.; Ståhlbom, B.; and Åkesson, B. Dimethylethylamine in mould core manufacturing: Exposure, metabolism, and biological monitoring. *Br. J. Ind. Med.* **1991**, *48*, 203–207.

27. Anders, M.W. (Ed.). *Bioactivation of foreign compounds*. Academic Press Inc., Orlando, USA, 1985.

28. Klaasen, C.; Amdur, M.; and Doull, J. *Casserett and doull's toxicology: The basic science of poisons*, 3rd ed. Macmillan Publishing Co, New York, USA, 1986.

29. Creasy, D.; Ford, G.; and Gray, T. The morphogenesis of cyclohexylamine induced testicular atrophy in the rat: In vivo and in vitro studies. *Exp. Molec. Path.* **1990**, *52*, 155–159.

30. Brust, K. Toxicity of aliphatic amines on the embryos of the zebrafish *Danio rerio*— Experimental studies and QSAR, Dresden, Dissertation. Technischen Universität, Dresden, Germany, 2001.

31. BUA [Beratergremium für umweltrelevante Altstoffe (BUA) der GDCh (Hrsg.), id est: The Advisory Committee on Existing Chemicals of Environmental Relevance of the German Chemical Society]. *Morpholin.* BUA-Stoffbericht 56, Germany, 1990.

32. Nelson, S.D. In *Bioactivation of foreign compounds*, M.W. Anders (Ed.). Academic Press, Inc., Orlando, USA, pp. 349–374, 1985.

33. Ramos, E.U.; Vaes, W.H.J.; Verhaar, H.J.M.; and Hermens, J.L.M. Polar narcosis: designing a suitable training set for QSAR studies. *Environ. Sci. & Pollut. Res.* **1997**, *4*(2), 83–90.

34. van Wezel, A.P. and Opperhuizen, A. Narcosis due to environmental pollutants in aquatic organisms: residue-based toxicity, mechanisms, and membrane burdens. *Crit. Rev. Toxicol.* **1995**, *25*, 255–279.

35. Bonse, G. and Metzler, M. *Biotransformation organischer Fremdsubstanzen*. Georg Thieme Verlag, Stuttgart, Germany, 1978.

36. Hypercube, Inc. (2002) HyperChem 7.01 [Program package].

Index

Milton Keynes UK
Ingram Content Group UK Ltd.
UKHW031148141024
449569UK00024B/987